韩国建筑世界出版社 编

155个居住设计 HOUSING 155

欧洲 EUROPE

美洲 AMERICA

大洋洲 OCEANIA

上 册

大连理工大学出版社

目录 Contents

出版人的话
Publisher's message

The lexical meaning of housing is a building that protects men from natural harms such as rain, wind, cold and heat and social harms such as theft and destruction. It is not an over-statement to say that human history had begun together with housing. Men cannot live without housing.

Being faithful to basic is always the most important. Therefore, I believe it is quite meaningful to collect the housing of the world, which is the foundation of architecture and the essence of living, and introduce them to the readers.

The five Olympic rings which symbolize five continents are the motif of this book. In the Olympic rings, the blue, yellow, black, green and red rings from left are overlapped in W-shape. Likewise, this book introduces the multi-family housing recently completed in Europe, America, Oceania and Asia. The concept of family is also changing now and this book tries to suggest the future direction of housing by showing the rapidly changing concepts on housing and residence by way of showing recently completed architectural works.

I sincerely hope that "I·HOUSING" will become one of the bases in learning the housing trend in Korea and overseas to the readers. It would be my pleasure if readers would get useful information and wise insight from this book.

As last, I would like to express my sincere appreciation to the architects and owners who helped us with information and photographing opportunities.

I promise that the Archiworld will continue to provide dear readers with solid contents and great editing. Meanwhile, your continuous interest and support would be highly appreciated.

JEONG, KWANG YOUNG
Publisher

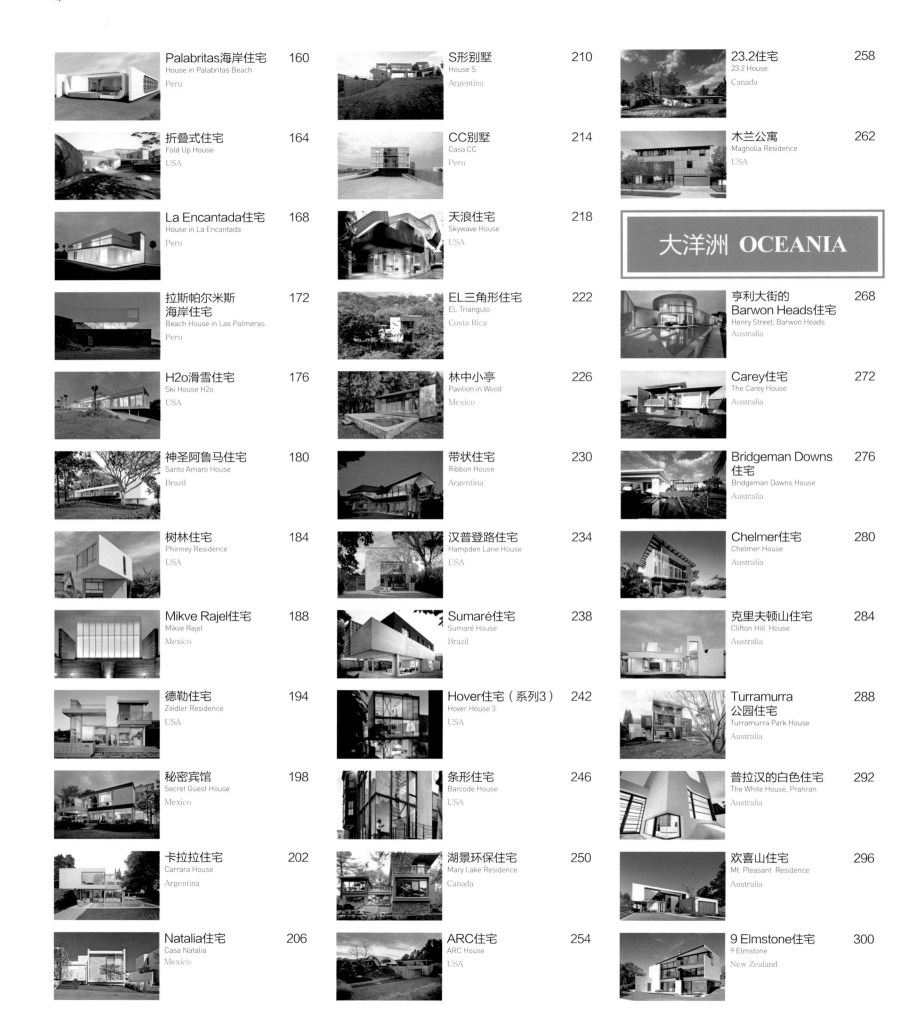

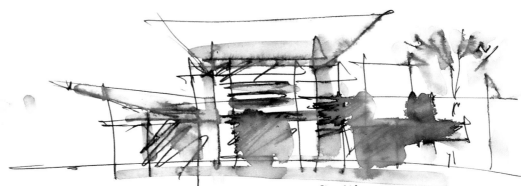

欧洲 EUROPE

格德拉普别墅 Villa Geldrop

Location Netherlands
Size 415㎡
Architecture design Hofman Dujardin
Architects, Amsterdam
Photography Matthijs van Roon,
Amsterdam

Hofman Dujardin Architects designed a large villa set in the Dutch countryside. It is not an ordinary house, however. What appears at first sight to be a simple block with an angled roof turns out to be a complex composition of space and light as well as a study of the modern home's function.

The building is placed at the rear of a large, flat site with the horizontal lines of its façade reflected in the pathway from the road.

The roof and the ground floor are both large, angular and dark blocks which are set off by large panes of glass that keep the geometry clean and the appearance uncluttered. Below grade, the story changes. At the front of the house, running parallel to the pathway, concrete steps lead down towards a basement-

level patio, opening to a glass hall-way/family room. On either side, this subterranean space is lined by bedrooms. At the rear of the house, the gesture is continued in the shape of a long, sloping ramp up into the garden.

Section

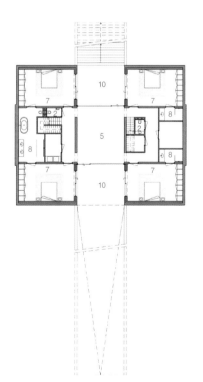
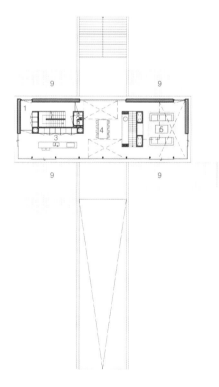

1. Hall
2. Toilet
3. Kitchen
4. Dining Room
5. Living Room
6. Study
7. Bedroom
8. Bathroom
9. Terrace
10. Patio

Floor plan

P字别墅 Villa P

Location Graz, Austria
Gross area 200㎡
Architecture design LOVE
architecture and urbanism
Photographer Jasmin Schuller

The site – a relatively steep hill – offers a stunning view over Graz. This panorama also provided the main guiding principle for the design. One additional goal was to provide direct access to the garden. Due to the very steep slope, this goal actually worked contrary to the aim of maximizing the panoramic view.

The upper level, which is oriented completely towards the fantastic view and contains all of the living space, while the lower level houses the doctor's office and adjoining rooms. This layout provides a clear separation of living and working spaces, including separate entrances.

The upper level features two spacious terraces – one facing south and west, which provides an extension of the living room and the children's rooms, and one facing north and east, which is an extension of the bedroom and bathroom and affords somewhat more privacy. A very broad external staircase with sitting steps connects the upper level and the garden, thereby joining the living space and the garden into an organic whole.

A pool is located in front of the building and connects to the carport. The pool and carport together provide a clear border between the property and the street, thereby increasing the privacy of the property.

Thus, a harmonious ensemble is created – a play of extrovert and introvert, of proximity and distance, which maximizes the strengths of the property, while simultaneously offering privacy.

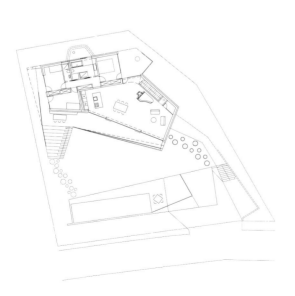

First floor plan

Site plan

十字小屋 CROSSBOX

Location Pont Péan, France
Surface 104㎡
Architecture design CGArchitectes
B3-Ecodesign
Photographer Javier Callejas

In the middle of a housing estate, a house is different : two crossing boxes cantilever over the other two. It's just surprising because of the straight volumes toped with greenery and the color.

This project is a prototype of a three-dimensional modular and industrialized house, built with four 40' shipping containers. The aim of this project is to build a low-cost architect's housing with high focus on environmental issues. With an industrial approach, the construction time is reduced, as the prices are getting down. Each volume presents a very simple design: living area on the ground floor, and three bedrooms on the first floor. The crossing of the two boxes provides a covered entrance and a carport.

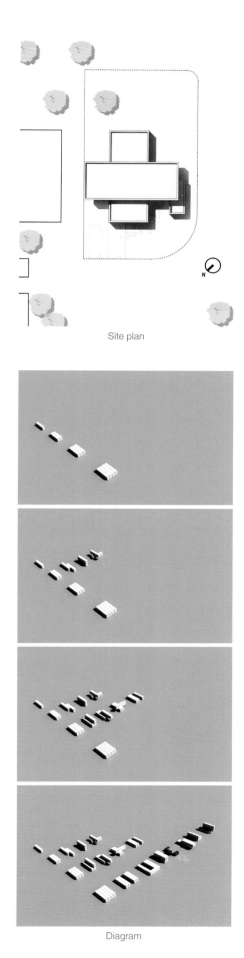

Site plan

Diagram

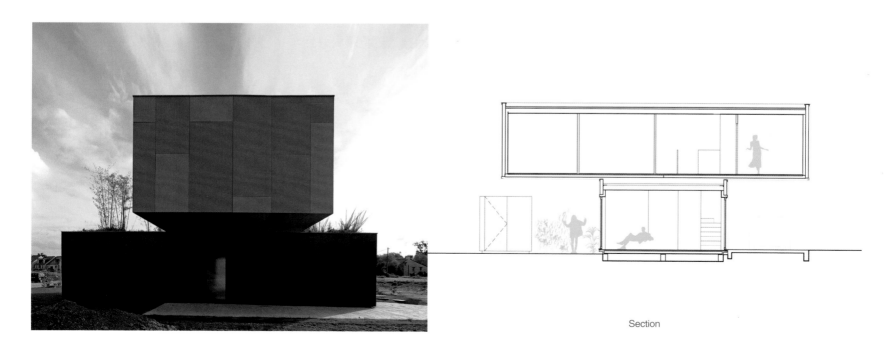

Section

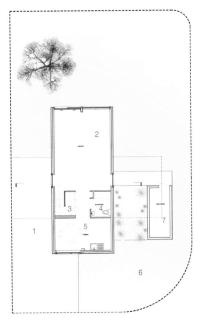

1. Parking
2. Living
3. Staircase
4. WC
5. Kitchen/Dining
6. Garden
7. Indoor Garden
8. Room
9. Bath

Farangas住宅 **Farangas House**

Location Paros Island, Greece
Site area 8,000㎡
Building area 420㎡
Architecture design React Architects /
Natasha Deliyianni, Yiorgos Spiridonos,
Theofanis Katapodis
Interior & Landscape design
REACT ARCHITECTS
Structural engineer
MFP consulting engineers
Constructional engineer Christofilakis
Nikos, Mechanical Engineer
Lighting study Mara Spentza, Architect
Photographer Dimitris Kalapodas
Collaborating architects Evi
Anastasiadou, Architect

The plot is located at the island of Paros in the Cyclades. It has orientation in the big side — south-western with view to the sea. The project concerns a summer house with hostels and a pool. The place of the residence is to benefit the view to the Aegean Sea. The house's volumes are placed on an "n" shaped layout with a central courtyard according to traditional prototypes. The central courtyard organises the operations, creating a central core in a direct report to the uses around it. Using the elements of the Cycladic Architecture, we designed a building that is adjusted to the landscape, split in different levels. In this way the adaptation of the building to the landscape is harmonious and the volume of the central residence is the least possible. The building from the eastern side is almost entirely adjusted to the ground. The architecture is subjugated in the landscape

and at the same time is using the intense Cycladic light to elect a particular morphology using as background of the arid landscape and the endless blue.

The house has references in the archetypal models of dwelling, with a modern certain output. In the interior the lack of bisector walls and the complete absence of elements such as curtains that would hide the light create a disposal of absolute freedom. Built by simple materials, plaster in white colour, floorings and bathrooms by poured material, the house is distinguished by austerity and simplicity in lines. The water and its light blue sense are presented with the swimming-pool, around which exterior spaces unfolds and views to the west and the sea. A space that submits its own pace, dictated by the light and air depending on the day and the season.

Site plan

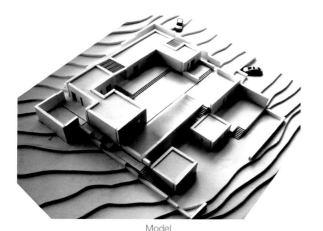

Model

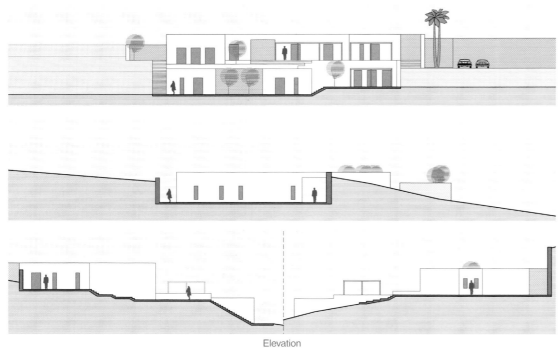

Elevation

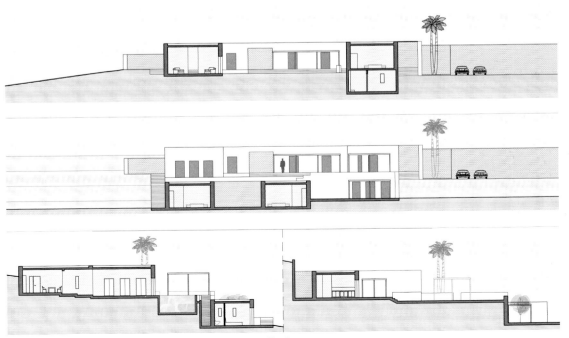

Section

1. Dining
2. Living
3. Bedroom
4. Kitchen & Dining
5. Bath
6. Sun Bed
7. Swimming Pool

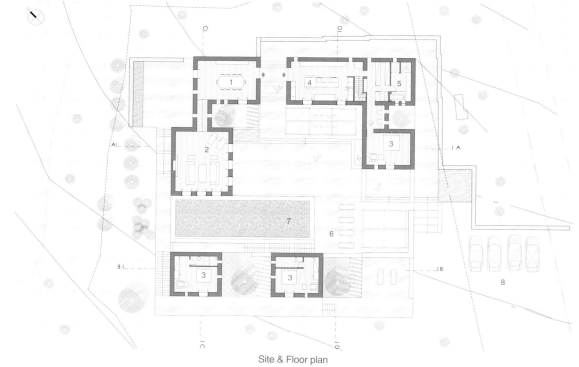

Site & Floor plan

博物馆大道别墅 Villa Museumlaan

Location Enschede, Netherlands
Architecture design SeARCH
Gross floor area 450㎡
Design participation Bjarne Mastenbroek,
Erik Workel
Assistants Pim van Oppenraaij,
Viviano
Villarreal, Maurice van der Steen,
Ton Gilissen
Photography Benno Thoma

The Museumlaan owns its name to the fact that it connects two museums, the National Museum Twente and Twentse Welle and the Cremer Museum.

Is drawing up an image-quality plan in which one refers to the cube-like architecture from the twenties and thirties of the last century still enough to create a consistent image?

At present, digital time films as "The Truman Show" and "The Martrix" show that images have become highly relative. There is such an inconceivable overload of images that the images itself have become useless. Isn't every house a simplified representation of living with elements as a living room, a garage, a kitchen or bedroom?

The design could be a "museum of daily life", a literal section of the house with all its functions clearly visible in the façade. Like this only a small section of the life of the residents is exposed without really exposing the life of the residents itself. The house is regarded as a dream or parallel universe.

How can one match maximum privacy to transparency, a nice contact with the garden, and an optimal relationship indoors and out?

The house stretches alongside the maximized allowed width

and height, therefore allowing the architectural element to be standing as a screen between the outside world and the residents.

It is possible, because of the linear realization of the house, that no function is condemned to a single sided orientation of, for example, the Museumlaan, because it is always stretched from the front to the back. Like this one has always the option to focus oneself completely to the garden, away from the street, and therefore the privacy can be enhanced. In the façade one can show with different materials or specific curtains the position of the different functions and like this, telling a story of its life.

The house as "museum of living" in which the curators can live in anonymity.

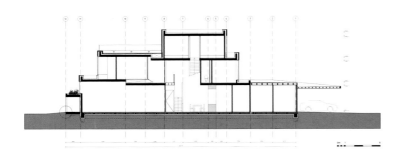

Section I

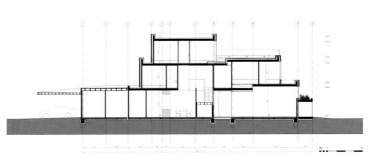

Section II

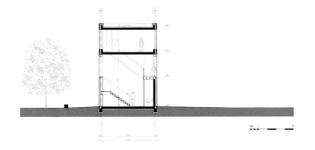

Section III

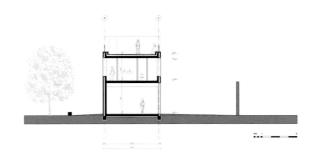

Section IV

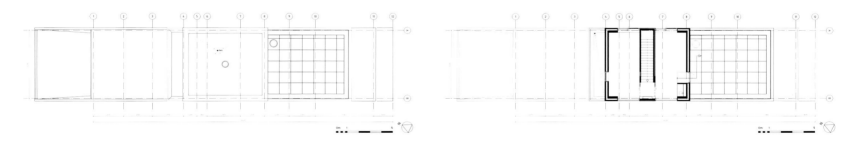

Roof plan

Second floor plan

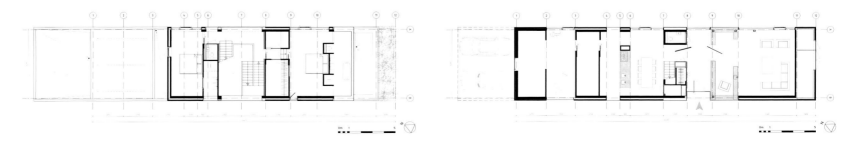

Frist floor plan

Ground floor plan

J字住宅 J House

Location Rocafort, Valencia, Spain
Architecture design(Design, Detailed Design and Site Management) bblab arquitectos (www.bblab.es), Ana Bonet Miró, Luca Brunelli
Design participation Jean-Baptiste Joye, Carla de Prada, Juan Lobato, Enrique Lopéz (Quantity Surveyor, Site Security and Management)
External consultants Jesús Egea, Structural engineer, Maria Ángeles González, Services
Photographer Ricardo Espinosa, (http://www.ricardoespinosa.es/)

A "wrapping" for everyday life.

How to combine an adequate privacy together with a straight relationship with the outdoor space in a small urban plot? How to enhance the spatial experience in a reduced residential program?

The building section articulates the public and private areas of the house. On the one hand, the sunny, open and transparent ground floor dissolves into the garden, whereas on the other hand a secretive, cagey and bright first floor introduces a more nuanced interior-exterior relationship. Here, the desired privacy is achieved by several patios enclosed by round-shaped lattice walls that allow seeing without being seen, and help to regulate the intense light of Valencia.

The hanging iron staircase acts as "transitional" device between the two floors. The spatial layout on both levels seeks to achieve further visual depth by constructing several interior-exterior sequences.

This is the project of a house characterized by its topological qualities rather than by its functional ones. The distinct and qualified spaces allow their residents to enjoy everyday experiences.

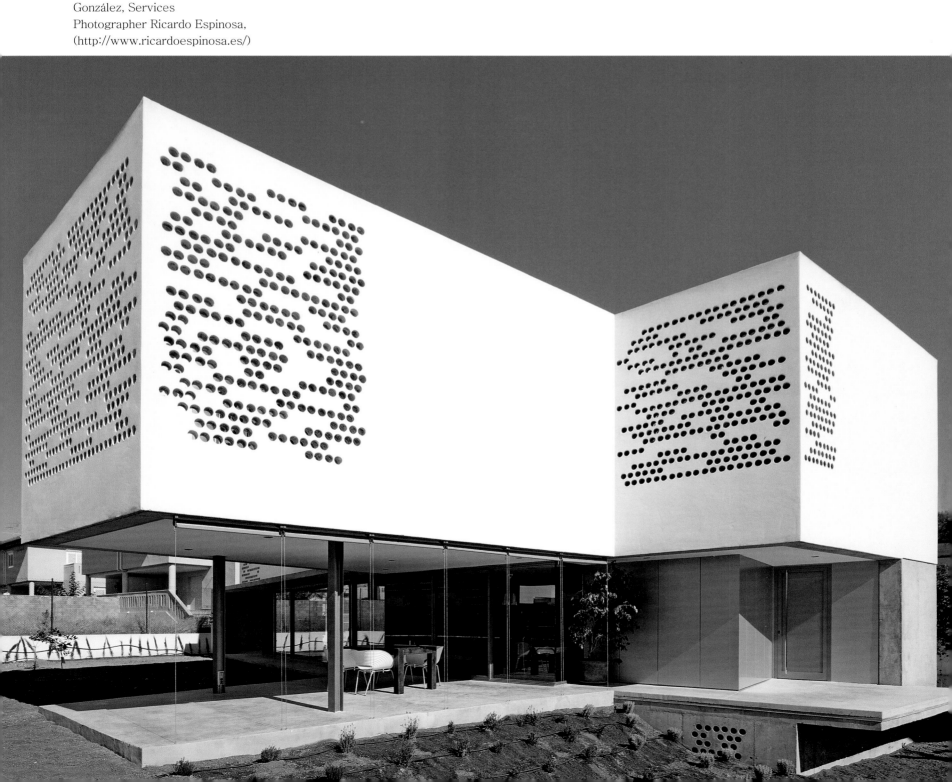

First floor plan and Cross section Ⅰ

Ground floor plan and Cross section Ⅱ

Basement floor plan and Cross section Ⅲ

3S别墅 Villa 3S

Location Graz Geidorf, Austria
Floor area 145㎡
Architecture design LOVE
architecture and urbanism
Collaborators construction
             calculation –
Hartmth Petschnigg
Building Physics – Roland Müller
Photographer Jasmin Schuller

A gorgeous property on the fringes of Graz Geidorf, a very limiting land-use plan and a very ambitious budget form the starting point for the planning of my own house.

Simple yet complex; clear but also playful; light and optimistic; small yet also big. A place that is architecturally distinct, yet eminently livable; unconventional and unique, yet still functional for everyday living – these are the attributes of my family's future home.

One of the fundamental ideas was to incorporate the relatively large property into the living space, meaning to make the boundaries between house and garden as fluid as possible in order to extend the living space over the entire property. This means as many subtle, boundaries and transitions between the inside and the outside as possible: large-scale vitrification with very large sliding doors; terraces that lead into the property and

sheltered areas serve to blur the borders.

The relatively strict land-use plan, which stipulated a saddle roof with a designated inclination and presented a further challenge. A folding begins at the seating platform on the southern terrace over the outside walls, continues over the roof, covers the building structure and thereby forms a "saddle roof" without taking on the appearance of a conventional saddle roof.

This folding spatially differentiates the individual areas, thereby providing more excitement to the entire complex. Due to the spatiality and "perspectivity" created in this way, the house looks different from every angle.

The interior rooms will be up to 4 meters in height, which will make the entire house appear much larger.

The interior of the house is centered around one main room for cooking, eating and living. When open, large sliding doors

between the individual rooms connect a fluid, complete spatial structure. With the sliding doors closed, each room maintains its intimacy and distinctiveness and also extends to its own outdoor area or access. For example, the bathroom has its own, opaque terrace with an outdoor shower, which can be converted to an interior room with the use of broad folding doors. Weather permitting, the bathroom space can be doubled in this way.

The building is a massive brick construction with concrete walls and a reinforced concrete roof. This roof also serves as a thermal storage mass to provide a pleasant indoor climate and is covered with wood outside. This shades the planking and visually blends the terrace and roof into a unified whole.

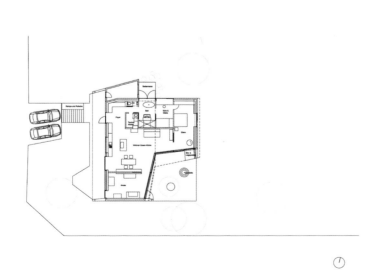

Ground floor plan

X5住宅 Casa x5

Location Treviso, Italy
Site area 620㎡
Bldg. area 620㎡
Total floor area 620㎡
Architecture design MZC
Architettura/Mario Marchetti,
Fabio Zampiero, Giuseppe
Cangialosi
Interior design MZC Architettura
Photographer Marco Zanta

A big apartment in the outskirts of the city, near the railway. The principal floor is for everyday functions. The upper floor is for the body, for relax, for playing. A central part is rounded by a black wall/furniture/wall. This black wall/furniture/wall is like a scenic curtain, a filter between day and night and vice versa. A big space living-kitchen is completely white: walls, furniture, ceiling, lamps etc. The flooring is like a water surface where white colour reflects itself. The interior concept is that every part has an individual life: every part contains space, light, colour and is composed by elementary volumes. All the surfaces reflect natural and artificial light giving to the space an overexposure effect. Crossing the black filter we enter to the night part of the house where the effect is similar to the day part. The difference between the two parts is the flooring: grey resin for day, cool grey wood board for night. Going up through the grey stairs we arrive on the upper floor. On one side we find the pool with a big wooden solarium around; on the other side a concrete-cladded kitchen with a concrete table.

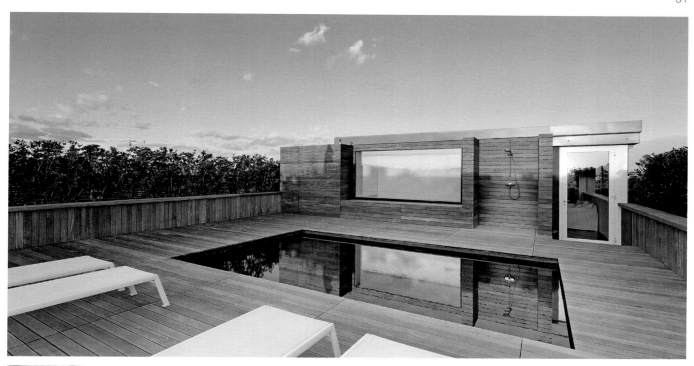

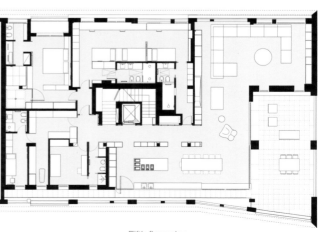

Fifth floor plan

▦ Detail

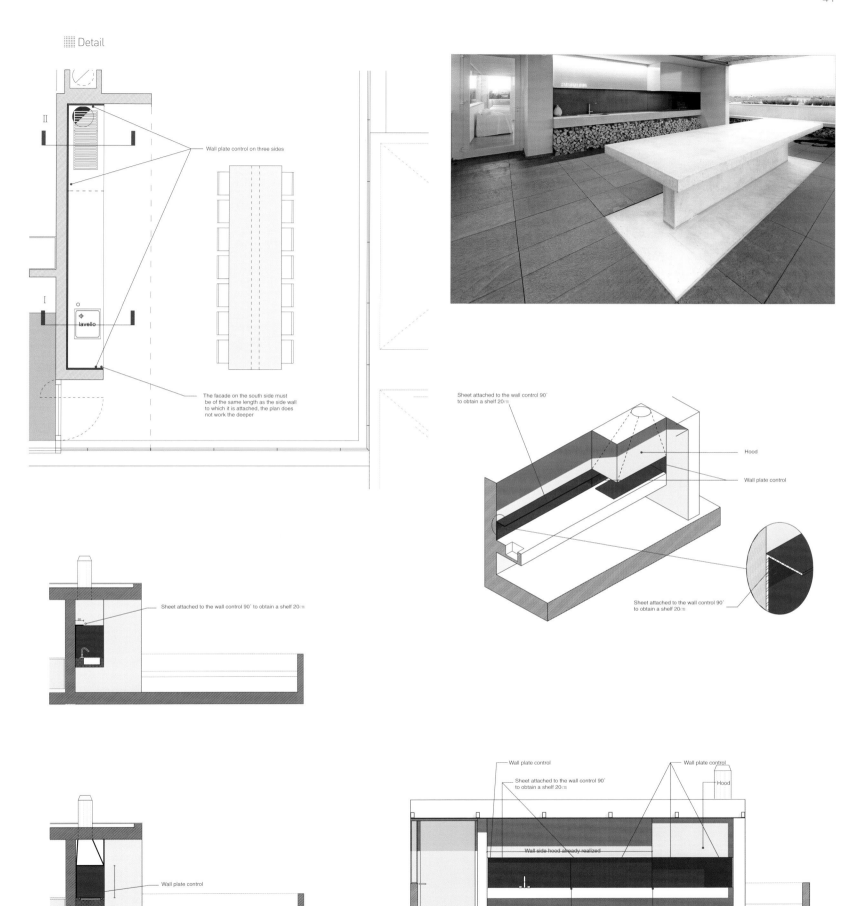

Wall plate control on three sides

lavello

The facade on the south side must be of the same length as the side wall to which it is attached, the plan does not work the deeper

Sheet attached to the wall control 90° to obtain a shelf 20㎝

Hood

Wall plate control

Sheet attached to the wall control 90° to obtain a shelf 20㎝

Sheet attached to the wall control 90° to obtain a shelf 20㎝

Wall plate control

Wall plate control

Wall plate control

Sheet attached to the wall control 90° to obtain a shelf 20㎝

Hood

Wall side hood already realized

Second division control wall plate in the middle

First division control wall plate flush with the hood control

Pustiměř别墅 Villa in Pustiměř

Location Pustiměř, Vyškov, South
        Moravia, Czech republic
Site area 1,209㎡
Built area 156㎡
Total floor area 173㎡
Architecture design studio NEW
WORK
Design participation M.A. Svatopluk
Sládeček
Collaboration Petra Bitaudeau
Vašková, Lucie Surá, Petr Janík

The neat little village is situated at the edge of the Haná plain where its southern slope raises to form the Drahanská vrchovina hills. The village was not damaged by any unfitting socialist construction in the 20th century or by any inappropriate interventions in the lawless years after 1989. Only now a new family house development is starting to grow on the village outskirts, luckily enough naturally extending and complementing the original street network. At the end of this area, under the ridge of the southern slope, the house in question is located. There is only a water tank and a game park above the house. There will be no more construction in this direction in the future.

The atmosphere of the place is reminiscent of a seaside resort.

The house gets lots of sunlight and overlooks a large stretch of landscape, is protected against northern winds and the agricultural plain is situated several dozen metres below the feet of an observer standing in front of the house.

The house is designed for two inhabitants with their children, grandchildren or guests. The bottom part is partly sunk into the ground and includes 2 bedrooms, a bathroom, a closet and a laundry room with a boiler room. The ceiling forms a continuously horizontal surface, a raised platform holding the main residential part of the house. The moderately broken flat roof rests on four pillars and the space is open to the exterior on the three sunlit sides with glazed walls. On the southern side there is the entrance to the terrace and the roof reaches

far behind the glazed wall to form a shielded and sheltered outdoor seating area protected against overheating. On the west the terrace level is connected to the ground with a ramp also covered with an extended roof. The windows in the western wall can also be shaded with a textile membrane. On the northern side, right on the ground level, i.e. three stairs lower, there is the garage, closing the main residential area on this side. The garage is passable so the necessary technology and materials can pass to the garden behind the house along the access road.

The simple geometrical ground plan of the house includes an oval capsule at the main entrance with a toilet for the upstairs part of the house. The downstairs level is entered via a staircase with a bookcase.

The house is built from ceramic masonry blocks. The ceiling and the roof are made of concrete. The house layout supports a favourable energy balance – the rooms for sleeping, where a lower temperature is usually required, are situated downstairs and the heat of the central heating naturally rises via the open staircase space to the upper residential floor. In the summer months the downstairs rooms are protected against overheating thanks to their position in the house.

1. Living
2. Entry
3. Hall
4. Storage
5. Garage
6. WC+Bath
7. Terrace
8. Bedroom
9. Bathroom
10. Service Room
11. Corridor
12. Clothes
13. Bedroom
14. Laundry

First floor plan

Second floor plan

Birkegade顶楼住宅
Birkegade Rooftop Penthouses

Location Copenhagen, Denmark
Size 900㎡
Architecture design JDS Architects
Client A/B Birkegade
Images @JDS Architects
Design participation Julien De Smedt,
Jeppe Ecklon, Sandra Fleischmann,
Kristoffer Harling, Francisco Villeda,
Janine Tüchsen, Claudius Lange,
Benny Jepsen, Andrew Griffin,
Aleksandra Kiszkielis, Nikolai
Sandvad, Emil Kazinski,
Bjarke Ingels, Mia Frederiksen,
Nanako Ishizuka, Thomas
Christoffersen, Eva Hviid
Morten Lamholdt

Optimization and utilization of the city roof

Elmegade district is probably one of the most densely populated areas of inner Nørrebro, CPH. Especially the triangular block Birkegade / Egegade / Elmegade has a very high density, which is reflected in very narrow courtyards.

And it is precisely around the cramped courtyard that the concept for BIR originates. The driving concept is to create the "missing garden" at the top of the existing housing block in association with 3 new penthouses, so all residents gain access to a genuine outdoor garden.

In order to qualify the "missing garden", JDSA looked at the Copenhagen gardens, which characteristically has an associated functionality. Therefore, a rooftop garden is designed as a space of functions and an associated materiality.

This is reflected in a playground with shock-absorbing surface and a playful suspension bridge, a green hill with varying accommodation backed by real grass and durant vegetation, a viewing platform, an outdoor kitchen and barbecue, and a more quiet wood deck.

The concept for the BIR, is to optimize and fully exploit the situations the site has to offer, and thereby design a potential for the future exploitation of the roof to the delight of all the co-

op's residents. It is a concept which is not limited to establish the 3 new apartments, but a concept which both creates a useful roof garden as well as a beautiful landscape for the co-op's neighbours and city residents in general. Usually a roof defines a final measure of any construction — closure. In the near future the Birkegade roof will open up for a versatile stay and experience.

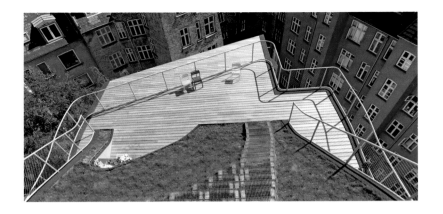

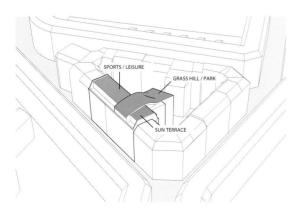

Diagram

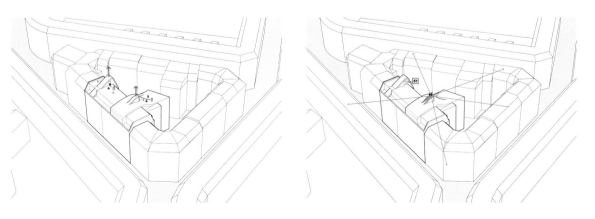

Diagram

阿姆斯特丹私人住宅 Private House in Amsterdam

Location Rieteiland Oost, IJburg,
Amsterdam
Site area 359㎡
Built area 280㎡
Client dhr. C. Carli
Architecture design Gert Jan
Knevel en John van de Weg
Design participation Jorrit Spel,
Giacomo Garziano, Bas van
Berkum
Project management Knevel
Architecten BV.
Contractor CEBO Groep
Advisor Adams bouwadviesbureau
BV.
Photographer Luuk Kramer

The residence consists of three floors, besides the basement and an entrance to the roof terrace. Its volume is following the plot's contours. The design of the building mass and the degree of "openness" determine the orientation of the house.

The entrance to the house is located at a small courtyard. From this side the house gives a closed impression. The façade has the maximum height permitted and acts as the back of the house, turning it to the courtyard. From this large façade the building mass slants downwards in one line to the South West. On that side the residence has an open character due to the use of large glass windows and the creation of loggias and roof terraces.

From the house the surrounding residential neighbourhood is hardly perceived because of its orientation towards nature. Therefore there is a strong sense of privacy and openness. The outdoor spaces on the upper floors, volumes which have been lifted out of the main structure, overlooking the surrounding nature as a result of the direction of the sloping roof. The many outdoor spaces on the floors also enhance the bond with the surrounding nature.

The building mass remains clearly readable as one volume by the continuous lines of the eaves/gables and facades.

A high level of abstraction is achieved in the facades by the composition of accurately executed openings and the selection of only two materials for the closed surfaces.

The upper part of the building mass is cladded with dark wooden profiles while the base at ground floor level is executed with light-colored rendered finish.

The abstraction is enhanced by fitting the complete sloping roof with anthracite solar panels which are well visible from the park across the water.

The solar panels are a vital part of the architectural image. Due to the almost identical dark colors of the cladding and the sloping roof and the accurate detailed connections, the roof and the facades create one volume. With the seeming simplicity of form, detail, materials and colors, the house is blending into its surrounding landscape.

Northwest elevation

Northeast elevation

Southwest elevation

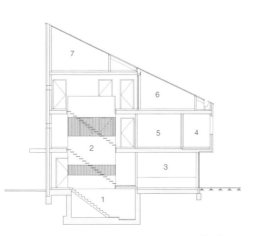

1. Basement
2. Staircase Wall
3. Kitchen and Living Room
4. Loggia
5. Master Bedroom
6. Terrace
7. Roof Terrace

Section

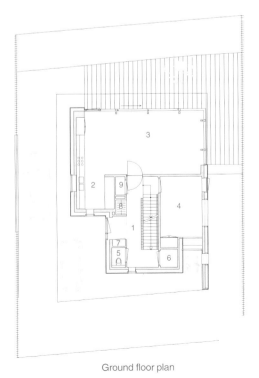

Ground floor plan

First floor plan

1. Entrance
2. Kitchen
3. Living Room
4. Study
5. Toilet
6. Storage
7. Technical
8. Wardrobe
9. Storage
10. Passage
11. Bedroom
12. Master Bedroom
13. Walk-in Closet
14. Storage
15. Toilet
16. Bathroom
17. Terrace

360住宅 360 House

Location Galapagar, Madrid, Spain
Site area 1,528㎡
Built area 385㎡
Architecture design
SUBARQUITECTURA
(Andrés Silanes + Fernando
Valderrama + Carlos Bañón) www.
subarquitectura.com
Structural design Subarquitectura
Building services Daniel Rodriguez
Client Arco Design and Projects
Photographs David Frutos Ruiz
info@davidfrutos.com

A unique opportunity for us in reality a problem that's been posed thousands of times: to construct a house with a public programme of social relation, associated with the private life of a numerous family on a sloping plot of the land with privileged views of the mountains outside Madrid. It has no one solution, there are many, they're even cataloged in books about houses of a slope.

We try not to think of domestic spaces. On the contrary, we take as a point of reference works of engineering, motorway intersections, changes of direction. We proceed from generic solutions to the problem of descending , solutions that conceal great plasticity. We seek the poetic in all that seems to have been considered from the merely pragmatic point of view.

The result is the literal construction of a used diagram. In this instance, form does not follow function, instead of function itself. Cyclical movement, routine and surprise turn into a way of living.

Its formal complexity offers the possibility of reaching all points of the house through two different routes, which multiply the possibilities of use and enjoyment. It has the form of a loop, 360º, like the shapes skaters make, like those of gymnasts, as artistic as they are precise.

An extreme shape, the house is curved, generating the greatest quantity of linear meters towards the good views. It is shored up in the landscape and turns back on itself, completing the revolution. The degree of intimacy increases as the distance to

the ends increases. At the midpoint, a mediatheque, isolated and completely dark, 100% technology, 0% landscape.

With a single gesture, two ways of moving are generated: going down and looking outwards. The long house, a sinuous movement, a descent by ramp and ample turning radii tangential to the setbacks of the land generate a panoramic vision. The short house, the quick way in a straight line, stairs of direct descent and a deep view towards the landscape.

A building that is black outside, absorbent, of slate, a material specific to the location, almost imposed as an aesthetic specification of the area. White inside, reflective, generic, neutral and luminous. Life incorporates colour, outside with the vegetation and inside with the people.

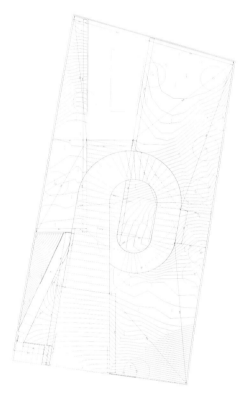

General plan

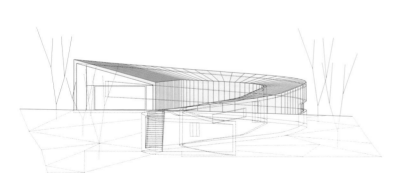

Elevation

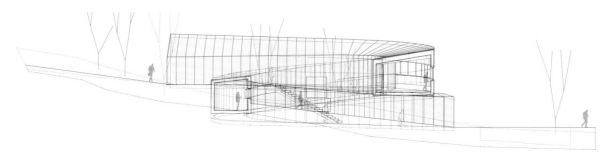

Section

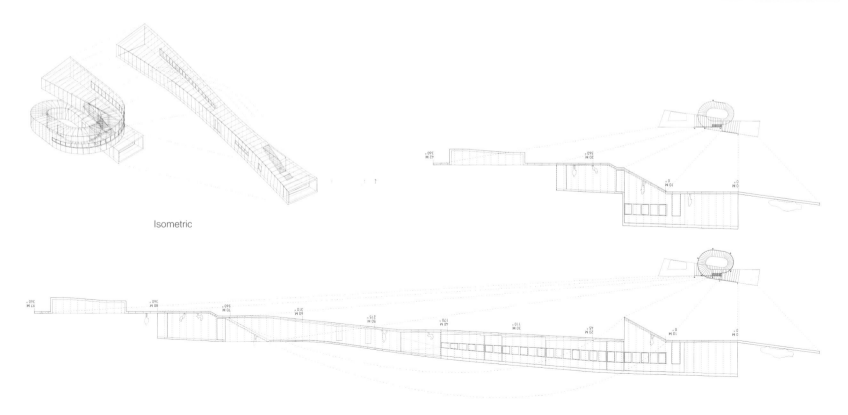

Isometric

Unfolded sections

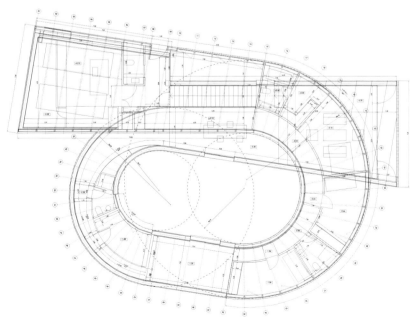

Floor plan

山腰住宅 House on Mountainside

Location Valencia, Spain
Site area 477,06㎡
Built area 230,00㎡
Architecture design FRAN
SILVESTRE ARQUITECTOS
Project architect Fran Silvestre,
Ma José Sáez
Interior design Alfaro Hofmann
Architectural technician Pedro
Vicente López López
Architecture José Ángel Ruíz Millo
José Vicente Miguel López
Fernando Usó Martín
Sara Sancho Ferreras
Photographer Fernando Alda

The building is located in a landscape of unique beauty, the result of a natural and evident growth. The mountain, topped by a castle, is covered by a blanket housing through a system of aggregation by simple juxtaposition of pieces generated fragmented target tissue that adapts to the topography.

The project proposes to integrate into the environment, respecting their strategies of adaptation to the environment and materials away from the mimesis that would lead to misleading historicism, and showing the time constructively to meet the requirements of the "new people". In this way the house is conceived as a piece placed on the ground, joining in the gap. A piece built on the same white lime, the same primacy of the massif on the opening, which takes the edge of the site to have their holes and integrated into the fragmentation of the environment.

The indoor space is divided by the void that is the core of communication cut parallel disposition of the mountain without touching it. On the ground floor are the garage and cellar, on a volume it has two floors with four rooms. Two of them, the rooms at the intermediate level are open to the private street, the other two on the upper level overlook above the houses opposite, the Valley of Ayora. One of them, the study is opened in turn to the central double height, incorporating it into their space. Across the gap, and on the mountain, are the areas facing the garden day illuminated by light reflected on the south slope of the castle oxidized.

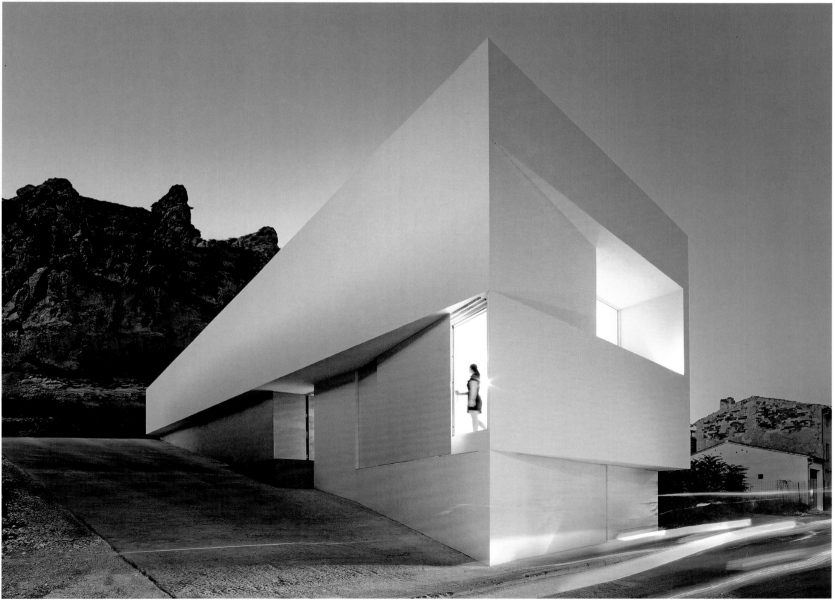

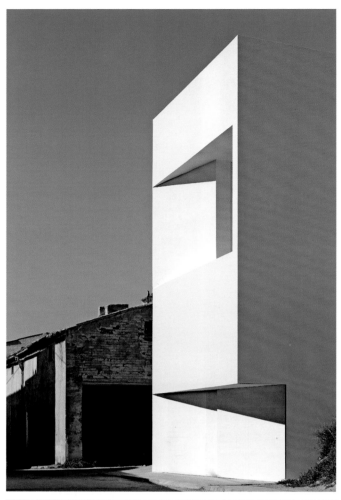

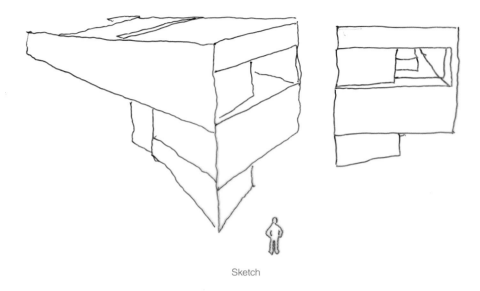

Sketch

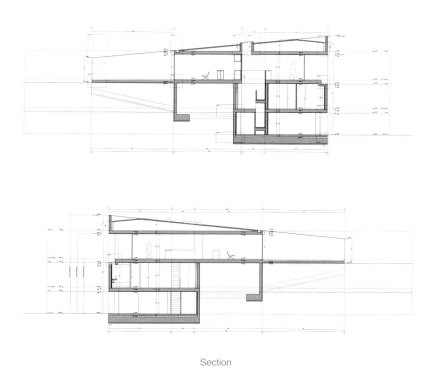

Section

First floor plan

1. Staircase
2. Interchange(Distributor)
3. Kitchen
4. Living Room
5. Study
6. Bath 1
7. Interchange
8. Bedroom 3
9. Terrace

Second floor plan

1. Staircase
2. Entrance Hall
3. Interchange
4. Bath
5. Bedroom 1
6. Bedroom 2

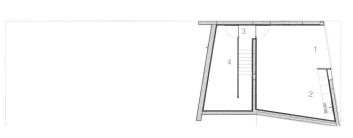

Roof plan

1. Garage
2. Laundry Room
3. Foyer
4. Bar

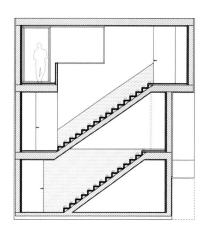
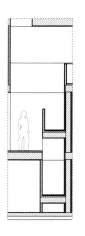
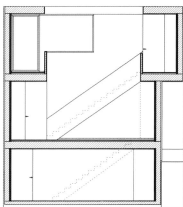
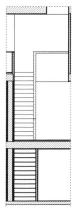

Detail (stair)

弹珠台住宅 Pinball House

Location Denmark
Bldg. area 508㎡
Architecture design CEBRA
Photographer Adam Moerk

The villa echoes its surroundings closely and opens up towards the magnificent view over the lakes of Silkeborg. The facade resembles a lifted skirt that floods the terrace and primary living areas in light. The broken line of the facade is continued into the central rooms of the house, where it looks like the ball in a pinball machine it shoots around to draw dynamic balconies inside its square volume.

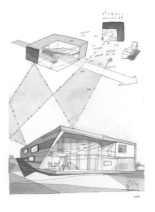

Sketch

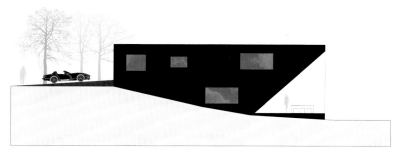

Elevation I

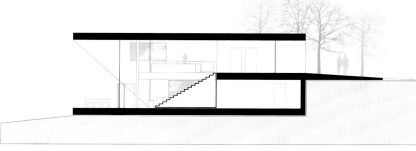

Section I

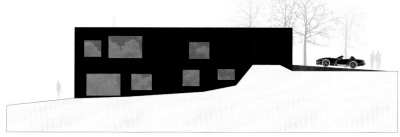

Elevation II

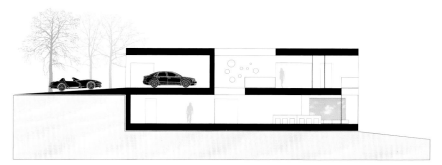

Section II

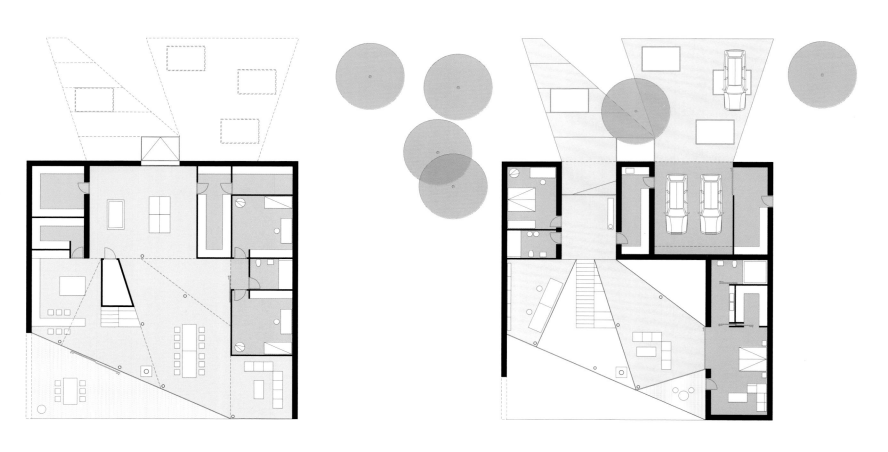

First floor plan

Second floor plan

C字住宅 House C

Location Barcelona, Spain
Bldg. area 398㎡
Architecture design RTA-Office
(Shanghai & Barcelona)(REAL TIME
ARCHITECTURE)
Santiago Parramon, founder-director
www.rta-office.com

The design of House C is characterized by the space, three-dimensional continuity, the skin and the cuts that let you discover the dermis. There are no elevations, no plans, or sections. This is space within a solid object, a block of black basalt in which we make a number of penetrations as in a quarry, mine tunnels and corridors that communicate with the outside, with the light of the morning, of noon and the darkness of the night.

An emotional design, the indivisible cross-sectional view of the object. Dark skin, pale dermis. Contrast. Cuts and cracks, reinforcing the change of scale, working for the presence. A unitary object that proposes its maximum dimension.

The complexity of the site works in our favour: the closed outer perimeter, walled, stoney; inside the glass folds open. We traverse the wall. It's the north face of the building and there we place the access: a spectacle of natural light, reflections, transparencies, an explosion of multiplying images.

A kaleidoscopic space, beauty, image, observation. All at the same time thanks to the effects produced by the transparent crystals of different sizes and angles. Everything impacts on the building. Nature enters the interior through these cuts and segments the different rooms of the house. At dusk, transparency overrides limits: the light of day is now projected from inside the house to the outdoors.

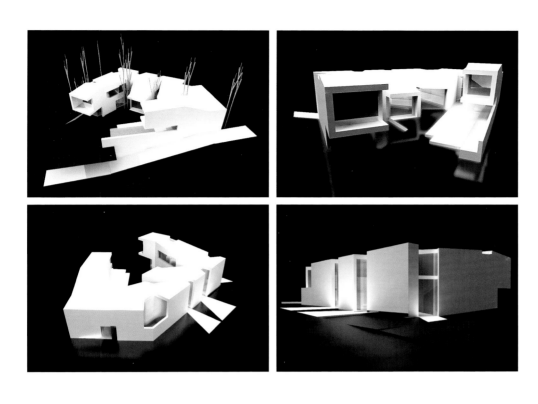

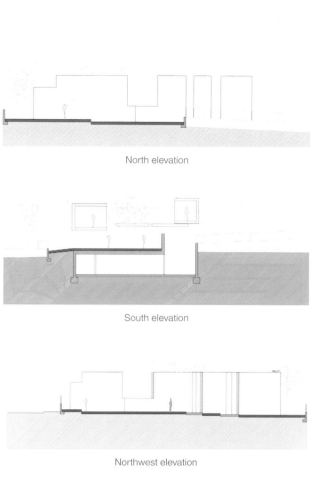

North elevation

South elevation

Northwest elevation

East elevation

沙丘住宅 Dune House

Location Thorpeness, Suffolk, UK
Architecture design Jarmund
Visgnaes Architects (JVA,
Norway) in collaboration with
Mole Architects(UK)

Mole Architects have collaborated with young Norwegian practice Jarmund Visgnaes Architects on this house. The house is situated on the edge of an area of outstanding natural beauty, overlooking the sea on the Suffolk coast. A complicated roof geometry draws inspiration from the seaside strip of houses with an eclectic range of gables and dormer windows. A robust design and access statement and extensive negotiations ensured planning success.

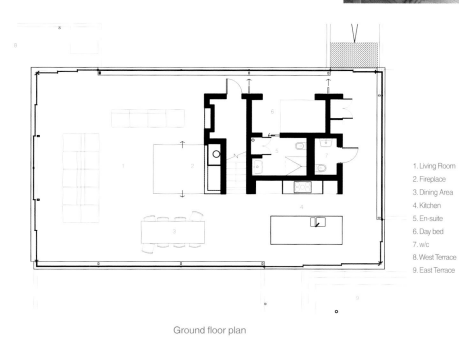

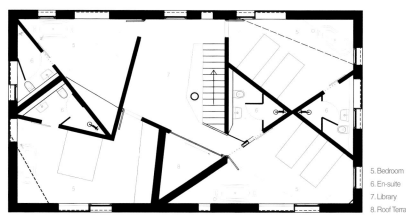

1. Living Room
2. Fireplace
3. Dining Area
4. Kitchen
5. En-suite
6. Day bed
7. w/c
8. West Terrace
9. East Terrace

5. Bedroom
6. En-suite
7. Library
8. Roof Terrace

Ground floor plan

First floor plan

SG Light住宅 House SG Light

Location Santo Tirso, Oporto,
Portugal
Site area 785㎡
Architecture design GRAU.ZERO /
Sérgio Nobre
Structural engineers Ricardo
Mendes, Eng.
Design participation Gustavo
Custódio
Photographer Manuel Correia

The idea blended a sculpture image with architectural needs, making them ambiguous, was the proposal made to the owner, as a work premise.

We used a 10° flexion in two volumes, in order to maintain among themselves a common language. The application of bending, runs one of the volumes in its horizontal axis, and the other one in its vertical axis. This gesture created a distant positioning to a formal level between the two block, allowing however have several readings volumes. These large volumes almost blind, providing you a very strong image.

Between the "bodies" happens the entry, where highlights, through a sectioning of the volume on the horizontal axis.

The organization of space, starts in the north wing of the house, a relationship between the house and an existing elementary school. Sought to break with some of the architectural forms of identification, especially outside windows that were used in the smallest possible number. Lighting/ventilation of rooms happens from individual small patios, that controls the temperature in a natural way without mechanisms.

The owners of the house SG Light are a young couple with interest and knowledge in contemporary architecture. As such, we had total freedom and support the decisions taken. The site shape is a trapezoidal rectangle. The site has an area of 785 m² and the house 240 m² of deployment.

The house is located in front of an elementary school, and children aroused the curiosity for the house and the architecture

in general. It was also well accepted by the neighborhood ... At first it was received with skepticism but was later welcomed.
This is a low budget home.

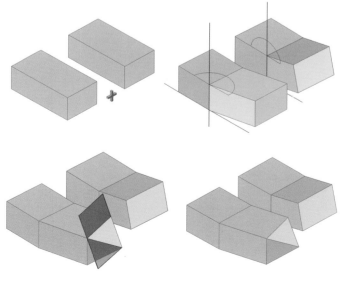

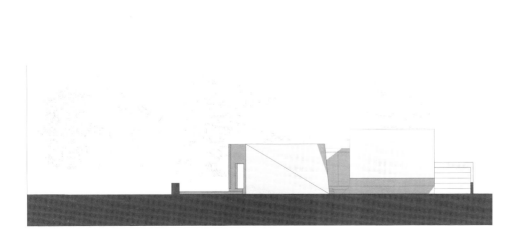

1. Bedroom
2. Bathroom
3. Kitchen
4. Living

Elevation

First floor plan

妮可丽娜山宅 Nikolina Gora Hill House

Location Moscow region, Russia
Built area 450m²
Architecture design za bor architects
Leading design & Principals Peter Zaytsev and Arseniy Borisenko
Photographer Peter Zaytsev

The name of the house is not incidental — "The House on Nikolina Gora" — which relates both to the geographical location of the house — near the elite village of Nikolina Gora (Rus: Hill of Nikolas), and the fact that the building is actually built into the hill.

The architectural design comes from the complexity of the landscape. The two-level lot allocated for the construction of the house has the elevation difference of 3.5 meters. The main entrance is on the second floor and the first floor overlooks a beautiful meadow.

Mostly hidden inside the hill, the two sections of the house —

a small and a large one — tower above the upper platform. In the small section, there are guest and staff apartments, technical facilities and the boiler. Public areas such as the living room, kitchen and the dining room are in the large section. Private zones are on the bottom floor, which is hidden from the road and other houses located higher on the hill.

The top public section is painted white. It is designed as a console over the private section and stretches wider on both sides by 3.5 meters. The private section is built into the slope of the hill. Such an arrangement required some specific design solutions, for example the beam-counter in the basement, sunk by 1.5 meters into the ground to support the massive console above.

Since the private section is in the hillside and the solar exposure is not optimal — in the central part there has been a clerestory cut, which adds to the volume on the upper level. The clerestory has a clever design and is equipped with the chute to drain the rain water onto the slopes where the plants are.

A similar clerestory is also on the roof. The roof has a minimal slope going toward the center of the roof, allowing rain water to be collected in the central part and flow down the inclined plane of the clerestory into the gutter and water the plants on the other side of the house.

This elegantly solves the problem of water-saving and improves solar exposure areas.

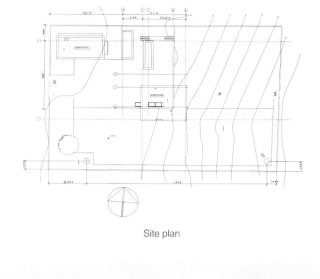

Site plan

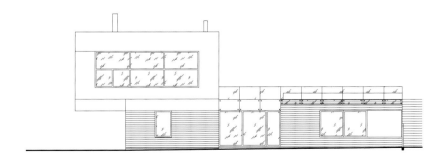

Elevation

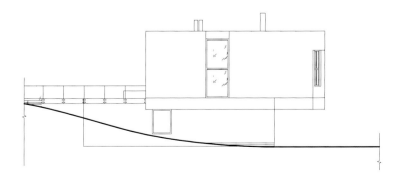

Elevation

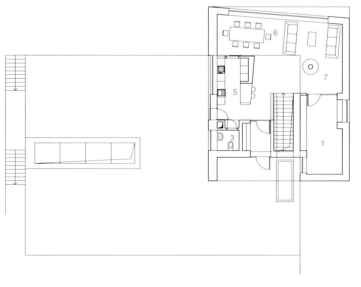

First floor plan

Second floor plan

J字别墅 **Villa J**

Location Danderyd, Stockholm
Site area 672 m^2
Designer Marge Arkitekter
Photographer Johan Fowelin

The client asked for a big family house for children to grow up in that could also host all relatives for dinner and harbour friends with broken hearts to heal. The site is on a beautiful hill with unspoiled nature and gleams of the sea in an old fashionable suburb of Stockholm.

By using a cross plan to organize the program we managed to divide the program in service, social and more private areas, yet keep them close. The cross also creates outdoor spaces that vary in character. A welcoming entrance yard, a sheltered place for the morning coffee, a garden with sun light all day and a back yard towards the north for storage of garden tools and wood. When moving around this big villa you pass different sequences as the facade changes its character. Windows varies in size from small apertures towards the north and the neighbour, larger openings towards east and totally glazed facade towards the southwest to gain the warmth of the sun.

The house is vertically connected in the middle of the cross by a staircase gallery in three stories, contributing to the passive ventilation. On the entrance floor the cross holds a service area with garage and kitchen entrance in one extension, the main entrance, a large social kitchen and a big living room in the other extension. On the second floor the extensions holds a service area with laundry and guest rooms in the first, children's bedrooms with playing area in the second and master's bedroom with bath and dressing in the third. On the third floor there is a relax area with sauna and roof terrace with views of the water.

Only durable materials are used such as brick, concrete, oak and copper that will grow old with patina. The facades are treated differently to accenture the different characters of the outdoor spaces.

1. Entrance Place
2. Backyard
3. Garden
4. Kitchen Yard

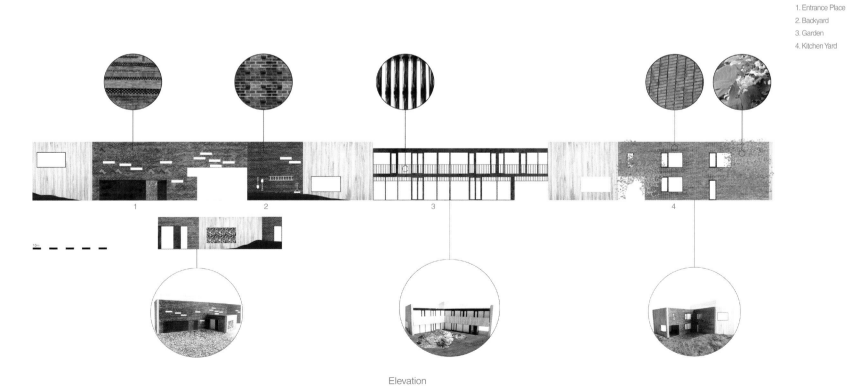

Elevation

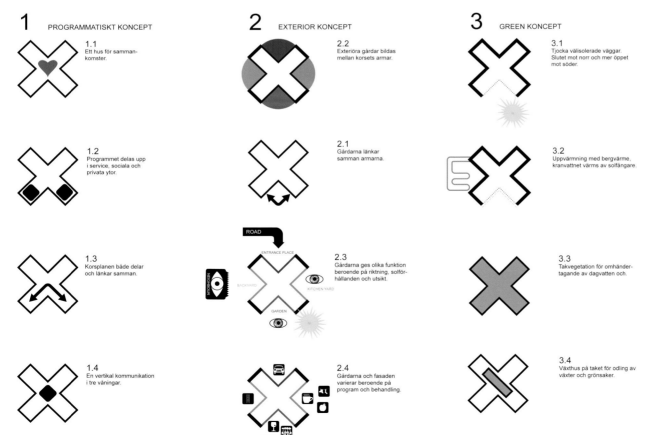

1 PROGRAMMATISKT KONCEPT

1.1
Ett hus för samman-
komster.

1.2
Programmet delas upp
i service, sociala och
privata ytor.

1.3
Korsplanen både delar
och länkar samman.

1.4
En vertikal kommunikation
i tre våningar.

2 EXTERIOR KONCEPT

2.2
Exteriöra gårdar bildas
mellan korsets armar.

2.1
Gårdarna länkar
samman armarna.

ROAD
ENTRANCE PLACE
BACKYARD   KITCHEN YARD
GARDEN

2.3
Gårdarna ges olika funktion
beroende på riktning, solför-
hållanden och utsikt.

2.4
Gårdarna och fasaden
varierar beroende på
program och behandling.

3 GREEN KONCEPT

3.1
Tjocka välisolerade väggar.
Slutet mot norr och mer öppet
mot söder.

3.2
Uppvärmning med bergvärme,
kranvattnet värms av solfångare.

3.3
Takvegetation för omhänder-
tagande av dagvatten och.

3.4
Växthus på taket för odling av
växter och grönsaker.

Concept

Section

1. Entrance hall
2. Stair case/Library
3. Living room
4. Kitchen
5. Patio
6. Garage
7. Stair case
8. Living room
9. TV-room
10. Bed room
11. Dressing room

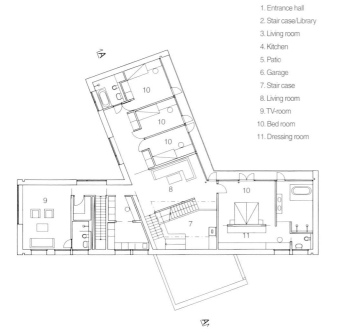

Second floor plan

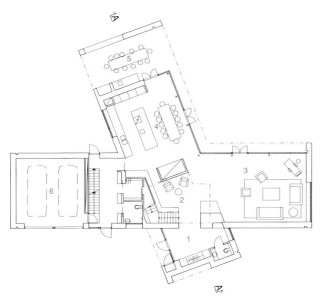

First floor plan

BH别墅 Villa BH

Location Zeeland, Netherlands
Site area 1,751㎡
Bldg. area 267㎡
Architecture design WHIM
architecture / Ramon Knoester
Photographer Sylvia Alonso

"Villa BH" is a modern,(environment)friendly house with a remarkable experience of space, light and the natural context. The villa is positioned on a rectangular plot of 35 x 50m, that is enclosed at 3 sides with similar plots and freestanding houses. On the back(north east) of the plot there's an old embankment with several tall trees, which existence is protected by local regulations. From the living program; the kitchen, dining area and living room are all orientated on this embankment with the large trees. Here the villa has a façade width of 20m. "Villa BH" is inhabited by a couple with 60+ of age. To optimize the accessibility of the house all the program is situated on the ground floor level around a patio. This enclosed outdoor space provides the owners of the house the privacy they admired. And

at the same time the patio makes the living area an enlightened space and gives it a façade to the south. On the other side of the patio is the main bedroom situated. By making the façade of the patio totally from glass panels, the main bedroom has a great view towards the existing embankment with several tall trees as a central focus point on the plot. The ceiling of the living area has an extra height in the shape of a sloped roof. The physical appearance of this area becomes hereby more specific and highly qualitative. Lifting the roof in this area also allows perspectives to the existing treetops, which give this plot its specific character, from all the different areas inside the building. The villa is designed as environmental friendly with extra insulated façades, with 30cm of insulation. With

this thick insulation there's a timber construction, that suits the thickness of the package. The roof is as well extra insulated and covered with sedum, which also regulates the distribution of the rainwater more gently. On the flat roof are 20 solar panels for electricity. A heat pump warms the interiors in the winter and cools them in the summer with natural temperature differences retrieved deep in the ground.

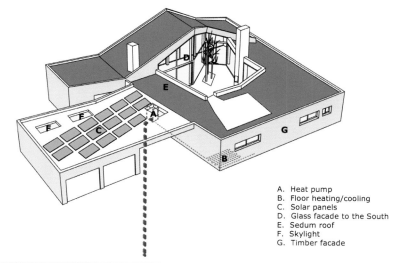

A. Heat pump
B. Floor heating/cooling
C. Solar panels
D. Glass facade to the South
E. Sedum roof
F. Skylight
G. Timber facade

■ Vertical detail North-East facade

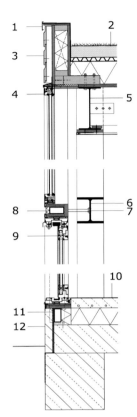

1. Sheet aluminium covering
2. Roof construction:
   .Substrate layer 80mm
   .Filter mat
   .Drainage 25mm
   .protective layer
   .Waterproof foil
   .Insulation 160mm
   .Vapour-retarding layer
   .Plywood 22mm
   .Structural beams 96x271
   .Acoustic ceiling
3. Siberian Larch tongued-and-grooved, waxed
4. Aluminium window frame with double insulated glass
5. Steel beam IPE 300
6. Steel beam HEB 180
7. Steel connection 15x15
8. Steel tube 100x50
9. Aluminium sliding door with double insulated glass
10. Floor:
    .Synthetic seamless floor finishing 3mm
    .70mm screed around underfloor heating
    .Insulation 120mm
    .Reinforced concrete slab 200mm
11. Steel tube 50x100
12. Prefab concrete

■ Horizontal detail corner North-East facade

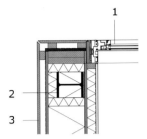

1. Aluminium sliding door with double insulated glass
2. Structural steel column, HEB 180
3. Wall:
   .Siberian Larch tongued-and-grooved, waxed
   .Cavity 38mm
   .Waterproof vapour transm. foil
   .Plywood 18mm
   .Insulation + wood construction 246mm
   .Plywood 18mm
   .Insulation 59mm
   .Vapour-retarding layer
   .Plasterboard 15mm

1. Carport
2. Entrance
3. Toilet
4. Installation room heat pump
5. Closet
6. Kitchen
7. Livingroom
8. TV room
9. Bedroom
10. Bathroom
11. Sauna
12. Guestroom
13. Patio
14. Garage
15. Terrace
16. Garden house

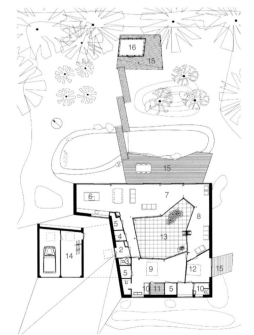

First floor plan

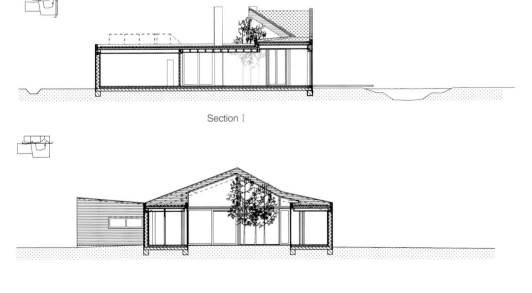

Section I

Section II

甲板之家 On the Deck into Life

Location Ljubno ob Savinji,
Slovenia
Bldg. area 420㎡
Architecture design SUPERFORM–
Marjan Poboljsaj, Anton zizek
Design participation Meta zebr
Constructional engineer PLUTON
Photographer Miran Kambic

The investor reconstructed two existing buildings. The houses are situated on the plateau beside the brook in the settlement Ljubno ob Savinji. The inspiration of the houses comes from the idea to create the atmosphere of the boat on the sea.

In the first house are bedrooms. It is meant for private, introverted ambient. The house is built from traditional materials. In the second house is a living room and a dining room. It is open, extroverted and is lifted from the ground level. It looks like a boat anchored in a green bay. The house is built from steel, glass, rheinzink, wood and slate. The interior ambientis designed as part of the architecture. Ambient shells are generic element, which through the repetition in different functional units offers us a rich sensual experience.

The exterior ambient is designed as fluctuation of the river. Paralel waves in front of the house create different ambients: wooden terraces, swimming pool, (fish) pond, area with high river grass, green plot.

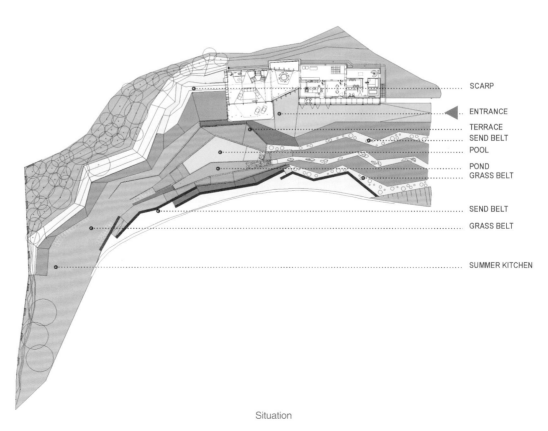

SCARP

ENTRANCE
TERRACE
SEND BELT
POOL
POND
GRASS BELT

SEND BELT

GRASS BELT

SUMMER KITCHEN

Situation

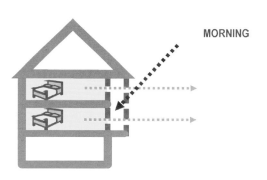

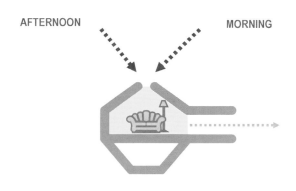

Eco - friendly diagram

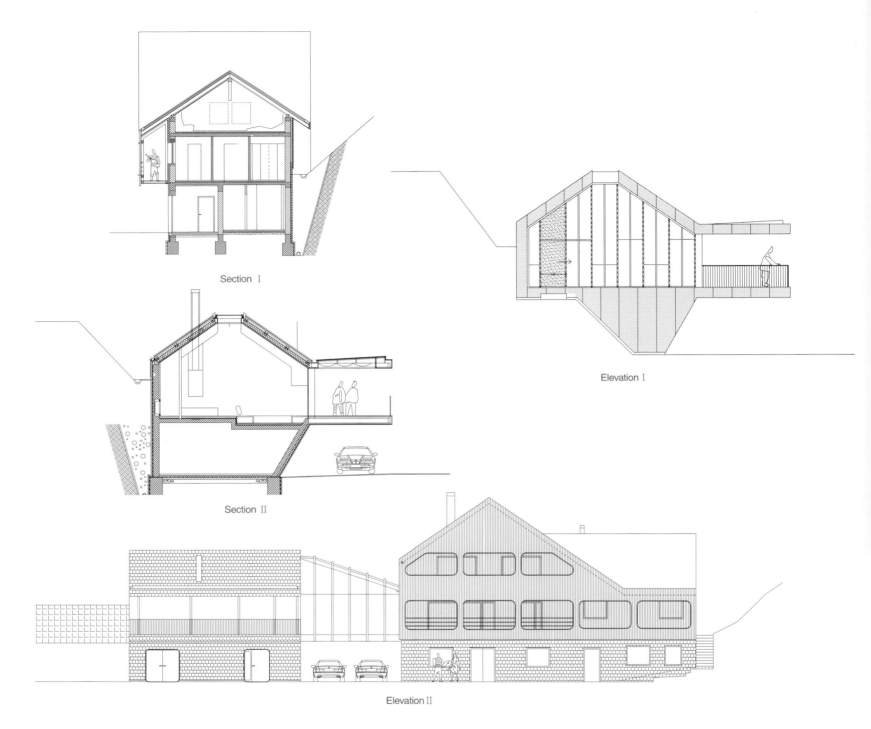

Section I

Elevation I

Section II

Elevation II

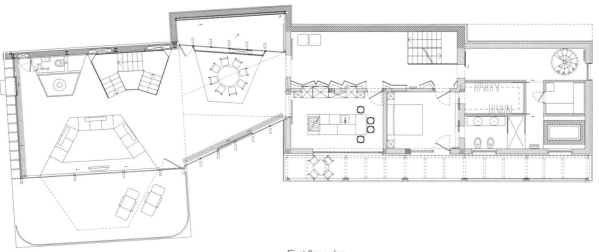

First floor plan

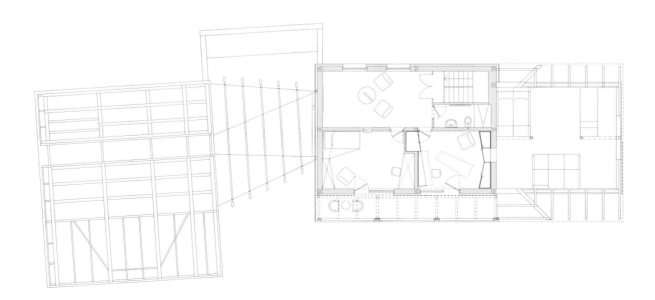

Second floor plan

Frenay别墅 Villa Frenay

Location Lelystad, Netherlands
Architecture design 70F
architecture /
Carina Nilsson
Design participation project leader_
Bas ten Brinke, engineer_van
Rossum Almere, contractor_ Ubink
Almere
Photographer Luuk Kramer

The commission consisted of a detached house with outbuildings on a plot in villa park "de Noordzoom" in Lelystad, at the foot of the Enkhuizen – Lelystad dike. The family wanted, like many others, light, air and space.

The house is a one-story bungalow, while the site asked for a response to "living at the water". The latter is accomplished by making a veranda along the entire length of the building (south), at the most important rooms of the house: from bedroom, to wardrobe, to the master bathroom, to hobby room, to the kitchen and finally to the dining room and around the corner the living room. Along this line, there is a transition from water to veranda, to complete glass facade, with doors behind which all

mentioned areas are.

To break the flat polder landscape, the roof of the living / dining room is slightly sloping. At the non-waterfront (north) of the house there are the children's rooms with their bathroom, a utility room and the living room. The existing artificial hills on the premises we accentuated, and the carport with storage room is half buried in one of them. The sauna with garden shed is placed on another hill, and is the highest of the three buildings. From the sauna there is a view over a Cor-ten steel waterfall that runs along the open-air terrace at the gable of the house, sloping towards the water feature that runs along the plot.

With the spread of the buildings, a residential area is created, more than a house with outbuildings. The transparency and direction can be seen in the interior, providing the ultimate inside - outside experience, emphasized by the slopes on the plot. The structural challenge of the T-shaped gable wall with glass corners, results in a phenomenal experience of the indoor and outdoor space in the living and dining room. The structural and architectural tension of the solution in this wall is experienced almost everywhere in and around the house.

1. Bedroom
2. Living room
3. Scullery
4. Workroom
5. Kitchen
6. Dining room

Floor plan

水池砖砌住宅 Brick Pool House

Location Chalfont St Giles,
Buckinghamshire, UK
Gross internal area 525 m²
(Basement: 100 m², Ground: 380 m²
(inc.link building): 45 m²)
Architecture design Paul+ O
Architects /
Paul Acland + Paulo Marto
Photography © Fernando Guerra

Paul+O Architects have completed a dramatic contemporary brick pool house in the grounds of a Victorian country house in Buckinghamshire. The new 525 m² building comprises a gym, playroom and a 15-metre swimming pool and features a link corridor to the main house.

The new addition to the estate is unequivocally contemporary in its design but its form and materiality take inspiration from the existing architecture.

Built of Belgian red brick with a steep pitched roof of handmade clay tiles, the new pool house presents itself as a contemporary annex which is sympathetic to the late Victorian red brick house and yet very much of its own time. The brick is slimmer than its traditional English red brick and a concealed gutter between the brick walls and clay tile roof gives the illusion of a continuous skin, contributing to the building's bold volumetric form.

Partners Paulo Marto and Paul Acland, who have completed a number of highly acclaimed residential projects, refer to the scheme as "playful and picturesque". At the corners of the building the robust brick walls give way to 3.8m tall Vitrocsa glass windows that slide back into the cavity wall, dissolving the boundary of inside and out and opening up the pool to the garden.

Strategically placed, the glazed openings relate to the house and the green houses of the adjacent walled garden and are placed to maximize winter sun and heat retention and prevent the pool from overheating in the summer. A long slim window at deck level frames a view of the walled garden from the pool so that swimmers can also enjoy the landscape.

The roof has a 30 degree pitch on two sides — which echoes with the steep roof of the main house — and a 17 degree on the other two, giving rise to an asymmetrical roofline which adds to the building's playfulness.

Internally the off-white Sto rendered walls and the Sto acoustic ceiling form one continuous skin, culminating in a full-length skylight over the pool (with opening section), which is set off-centre to flood the darkest part of the pool with natural light. Ventilation and services are carefully concealed and supplied via unobtrusive wall grilles and slots in the basalt stone floor.

The pool hall is heated by a heat recovery air handling system and the building achieves a good thermal performance through the specification of 300mm reinforced concrete structural walls with insulated cavity.

The swimming pool is connected to the house by a glazed corridor, from which one can also access the carport, a store and the changing facilities. The roof of the link corridor features a full-length skylight, which softly washes the brick back wall with natural light. At night a continuous ribbon of concealed fluorescents washes the same wall with artificial light. The link corridor provides access to the house and also functions as a sheltered dining/seating area with glazed doors, which open out on to the garden. Planters filled with Trachelospermum Jasminoides will eventually cover the entire back wall with green foliage and fragrant flowers, further enhancing the relationship between the pool house and its surrounding garden.

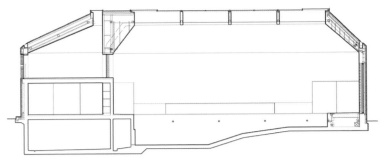

Section

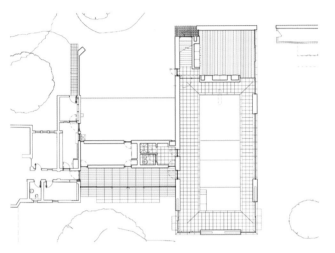

Ground floor plan

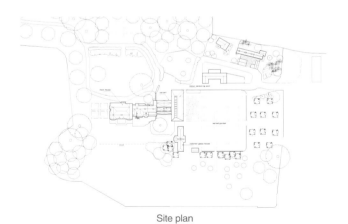

Site plan

Elevation

South elevation

North elevation

内面外向式小屋 Cabin Inside-Out

Location Ostfold, Norway
Site area 800㎡
Bldg. area 90㎡
Total floor area 80㎡
Architecture Design EIULF
RAMSTAD
ARKITEKTER
Design participation Anders
Tjønneland
Photographer Roberttodi Trani, Kim
Mllüer

The house is beautifully situated on the top of a hill overlooking the ocean and the horizon. It is placed in the midst of an uncultivated landscape on a small island. The small scale of the house, together with the use of wooden materials that will gradually develop a grey patina, allows it to coalesce with the existing forms and natural colors of the surrounding landscape. The design and spatial layout engenders a sensation of being outdoors while inside.

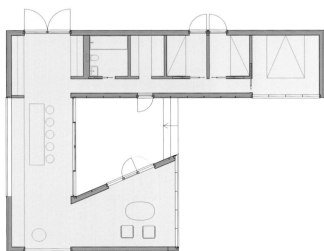

First floor plan

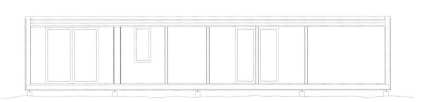

Section

Pavilniai地区公园的家庭住宅
Family House in Pavilniai Regional Park

Location Vilnius, Lithuania
Site area 1,647㎡
Bldg. area 327㎡
Architecture design JSC Architects
bureau G.
Natkevicius and partners
Design participation G.Natkevičius,
R.Adomaitis, R.Babrauskas,T.Kuleša
Photographer R.Urbakavičius

The custumer is a banker and an antique book collector. A four member family house. In the Middle Ages the area, where the building is situated, was a cannon foundry. Customers bought a site where stood the old yellow brick lodge with a basement. Cleaning the plaster of a house revealed that the lodge had been built by ancient bricks which were made in a old Vilnius brick factories. Becouse of a historical and physical value of a house were considered to preserve it by wrapping it with outer glass shape. Historical house structure have been carefully restored.

Library of a collection of ancient books equipped in the basement of historical lodge. On the ground floor are childrens' bedrooms and in the attic are master bedrooms. At the glass shape zone in the basment we can find the turkish bath with a rest rooms and a garage for two cars. On the ground floor there are living room, kitchen, dining room and a wardrobe. Glass form is not an end to itself. From each point on the ground floor area on a 360 degrees offers fantastic views of regional park. Ground floor space is like a yard of a historical lodge.

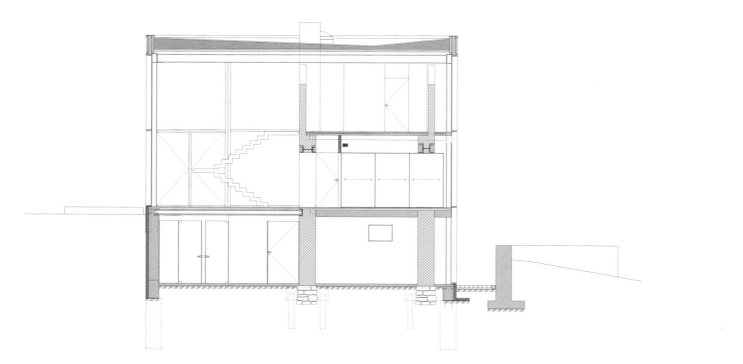

Section

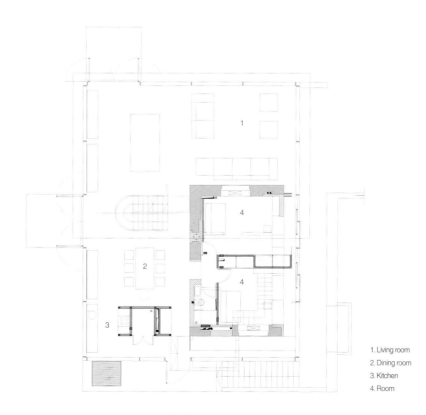

1. Living room
2. Dining room
3. Kitchen
4. Room

First floor plan

遮蔽式住宅 The Shadow House

Location London, UK
Total floor area 79㎡
Architecture design Liddicoat &
Goldhill LLP
Design & Client David Liddicoat &
Sophie Goldhill
Structural engineer Andy Martin of
Peter Kelsey Associates
Photographer Keith Collie & Tom
Gildon

The Shadow House is a new build private house in Camden, North London. We carried out the entire project ourselves, from finding the site, through planning, design, construction and manufacture of the fittings, fixtures and furniture.

The tectonic of the house was driven by its physical and legislative context. We developed a monolithic design, reinforced with unforgiving, monochromatic surfaces. Because our budget was so tight, we carried out most of the work as possible ourselves and limited our palette to primary materials.

The project's name (a conscious reference to Junichiro Tanizaki's poetic essay, In Praise of Shadows) comes from its skin, which is expressed in a black slim-format glazed brick. Integrated furniture, joinery and structure is made from various treatments of laminated larch. The interior is tuned to create moments of distinct character: with limited space available, we sought intense contrast between light and shade, different material textures and floor & ceiling levels to modulate the atmosphere.

The intense atmosphere of the living spaces contrasts with the dazzlingly bright top-lit first floor bathroom. We created a frameless glass ceiling to give the sensation of being outside; showering in full sunshine or bathing under the stars.

One small luxury we allowed was to buy two slabs of bookmatched Statuarietto marble, which we used throughout house as a reflective contrast to the brick walls. The whole design revolved around this play of light & dark, carefully controlled moments of intensity and quiet shadow. We wanted to create interior spaces with maximum emotional effect.

Just building a house doesn't make a home: we also designed our fittings and furnishings; the minimalist Zero larch bedframe; kitchen cabinetry in elm, stainless steel, marble and spray lacquered matt doors; The Shadow Lamp, a granite and laser-cut timber light; soft furnishings using amazing African fabrics, Nyaradza bedspread and Akwasidee cushions.

1. Bin store
2. Bike store
3. Front entrance
4. Front entrance courtyard
5. Entrance
6. Snug
7. Stairs to first floor
8. Dungeon

9. W.C.
10. Kitchen
11. Dining room
12. Rear courtyard
13. Landing with glass roof
14. Master bedroom
15. Library
16. Bathroom with glass roof

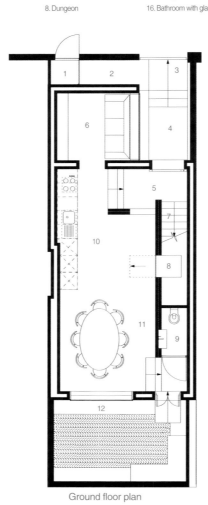

Ground floor plan

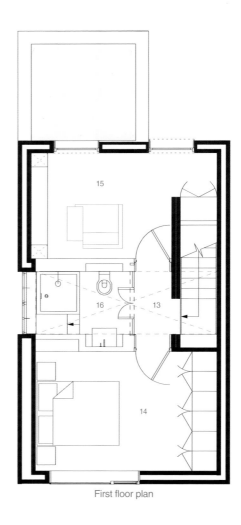

First floor plan

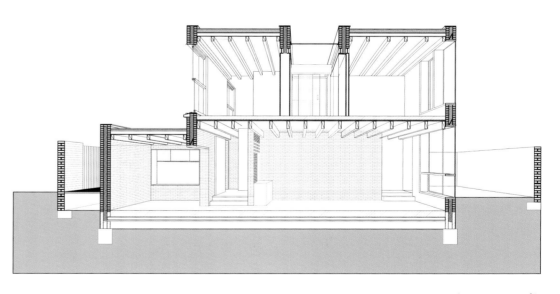

Section

0 ____ ____ 5m

Bateman's Row住宅 **Bateman's Row**

Location London, UK
Architecture design Theis + Khan
Cost consultant –
Stephen Cuddy
Structural Engineer –
FJ Samuely and Partners
Lighting Design –
George Sexton Associates –
Planning Consultant –
CMA Planning
Planting –
James Catoe Design
Photographer Nick Kane nick@
nickkane.co.uk

TKA have created a mixed-use building comprising three floors of commercial space, with four residential units. The design has utilised a confined site to create a contextual building within the existing cityscape. There is a deliberate play of opposites throughout; flush and recessed, rough and smooth, dark and light, solid and void within an overall carefully proportioned composition that develops from a dark solid based to an open glazed top.

Sustainable construction methods include an exposed concrete structure for thermal mass combined with a highly insulated envelope; solar panels supplement hot water provision; natural ventilation and a green roof.

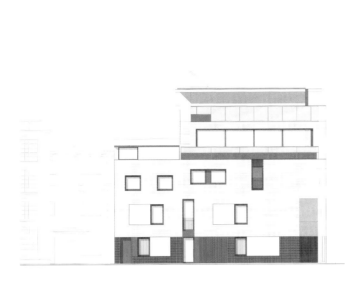

West elevation

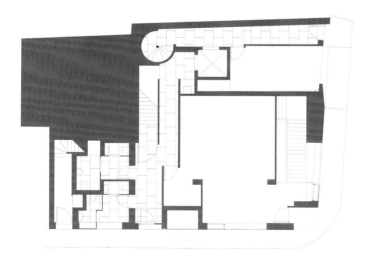

Ground floor plan

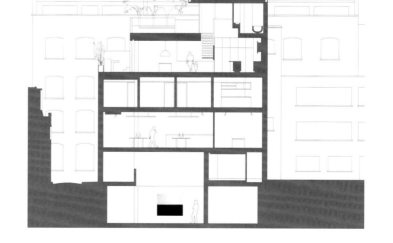

Section

Charrat住宅 House in Charrat

Location Charrat, Valais,
Switzerland
Gross floor area 230㎡ and 740㎥
Architecture design clavienrossier
architectes hes/sia
Design participation Valéry Clavien
and
Nicolas Rossier
Photographer Roger Frei
http://www.rogerfrei.com

Situated away from a small village in the middle of the Swiss Alps (Canton of Valais), this 19th century house included an adjacent barn and had a too vast volume to be renewed in its totality.

The existing building had no special qualities apart from the thick stone walls hidden by external plaster cement. The double-sided pitched roof was too low to allow us seeing the surrounding landscape from the garret floor. The existing windows were small; a large part of the volume was blind, half the building was being used as a barn. Consequently, we preferred to intervene directly on the existing building, walls and roof included.

We kept what was useful for the project, cellars, first floor and half of the second floor of the old house, demolished the remaining elements and rebuilt according to new rules that was made especially for the project. We wanted to create an ensemble that could communicate with its environment, vineyards, stone walls and the Alps. This way of intervening was as well a response to the client, not to transform the totality of the existing building, cost-wise. We were able to reduce the existing 320 square meters down to 230 square meters.

We wanted to create a strong contrast between the remaining part and the new structures. We chose to oppose clear geometric lines with existing rough old stone walls. Volumes of visible tinted concrete replaced the double-sided roof and the transformed area. We tried to create an ensemble, to establish

a dialogue with the existing. That was the reason why we opted for a massive construction made in reinforced concrete, with a stone-like color. The oxides added in the concrete made the hue similar to a tuff, stone found in a very small quantity in the stone walls. Both of the new concrete volumes are sitting atop sixty centimetres width existing wall. By adding insulation and lining, you reach eighty centimetres tick. This thickness issue has motivated us to work on the wall depth and became as well a response to a formal desire. Thus, the idea of sloped walls were chosen to erase, at least visually, the thickness of wall; to be opened outwards while maintaining a solid appearance with the existing structure. The various-slopes faces enhance the highly varied game of the shadows throughout the day.

The openings in the existing part were small and vertical. We kept them to accentuate the contract with the newly created on top. We made large horizontal windows, thus becoming frames on the landscape. We chose to make a single front opening on each concrete facade. This desire of openness is also visible in plan. Interior walls are not touching the facades. This system allows us to experiment in each room a transversal views onto the landscape.
There are no corridors. Circulation is made along the external wall, from room to room. The overall view continues beyond the windows, opening onto the surrounding landscape.

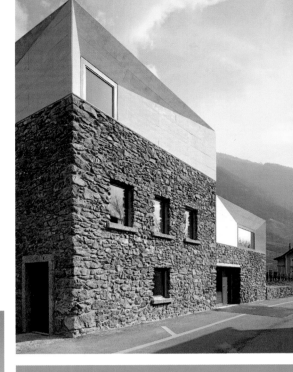

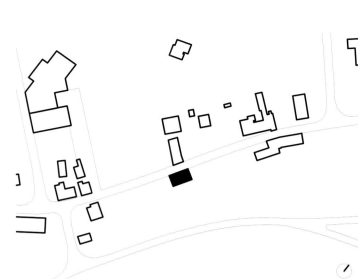

Site plan

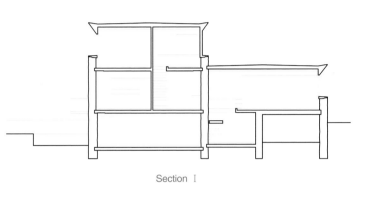

Section I

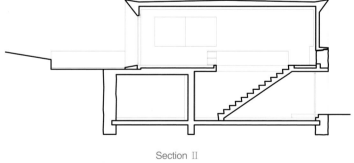

Section II

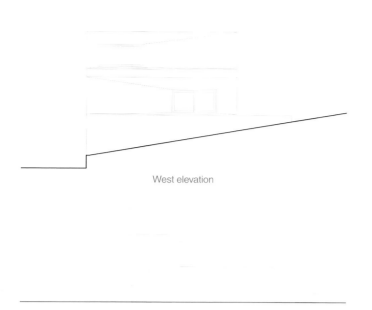

West elevation

North elevation

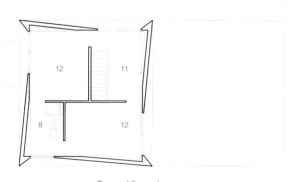

Second floor plan

1. Entrance
2. Cellar
3. Laundry
4. Carnotzet
5. Technical Room
6. Living Room
7. Kitchen
8. Bathroom
9. Master Bedroom
10. Terrace
11. Internet / Ironing
12. Bedroom

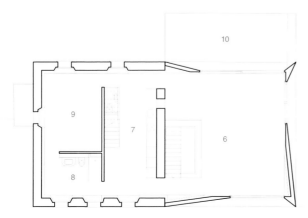

First floor plan

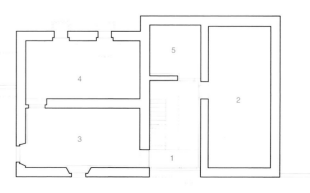

Ground floor plan

77住宅 House 77

Location Póvoa de Varzim,
Portugal
Constructed area 232㎡
Architecture design dIONISO LAB
(www.dionisolab.com)
Desingn José Cadilhe
Design participation José Cadilhe,
Emanuel Fontoura (Final Design)
Contractor Consarte Lda.
Photographer Fernando Guerra

Póvoa de Varzim is a city profoundly related to the sea and fishing. Its great cultural richness became an interesting stimulus to the project. In fact, the house was an opportunity to revitalize some of the city's memories and to participate in the panoply of colours and materials that characterise the street.

The house is simple, it is organized in a vertical and hierarchical way. The social areas are on the inferior floors and the private areas on the superior levels. To achieve great visual amplitudes and dynamic interconnections between spaces, the interior was structured in half floors. The width of the plot decided the stair. In fact, it became the heart of the house. A wall painted with Blue Klein emphasizes its importance and continuity through the spaces.

The west facade is covered by aluminum venetian blinds that not only defend the interior from the insulation but also open the house to a small garden. At east, the house gets its identity. The intimacy is guaranteed by stainless steel panels, perforated with the "siglas poveiras". These symbols are a proto-writing system once used as a way of communication and to mark personal and fishing belongings. Also, they were hereditary and constituted an important family legacy that was transmitted by inheritance through generations, evolving with new combinations.

In this way, the house, in the very centre of "Bairro Norte", shares some of the city's memories and references with the population and revitalizes a legacy that has been progressively forgotten and abandoned. Quietly, the house confesses its pride in the city.

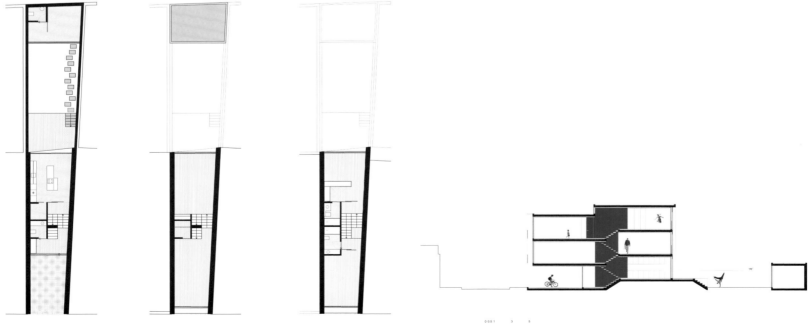

Ground floor plan   First floor plan   Second floor plan   Transversal section

恩波利住宅 House in Empoli

Location Empoli, Florence, Italy
Total area 220m²
Architecture design Massimo
Mariani
Design participation Elda Bellone,
Roseda Gentile, Alessandro
Mariani
Photographer Alessandro Ciampi

The Stefano's house is located in Navicelli road, between the historical center of Empoli and the Arno river. The project concerned the renovation of a residential building which was built in the earliest of 1960's. The plan was to empty entirely the interiors so we could reorganize the interior floors height; in the same time we designed the new façade on the main front without any increase of volume.

The building is developed on three floors above the ground; from a monofamiliar house it turned into a bifamiliar house, with a few offices on the ground floor. Each new house measures 200/220 square meters about, both the two have the entrance door along the main road as the garage and a service room. The first floor holds the living room, the dining room and the kitchen, each one

offers many different views. There is a terrace along the road which is shelded by an aluminium wall with laser pierced flowers; on the other side the kitchen and the dining room overlook on the interior patio with the hydromassage pool. In summer it can be closed on the top by the motorised curtain-blind so it turns into another room. A teak staircase connects the living room with the night zone on the upper, floor which is opened on the river on a side and the city's roofs on the other. A small green house amuses the home entrance.

1. Kitchen
2. Dining Room
3. Living
4. Patio
5. Bath
6. Terrace
7. Indoor Terrace

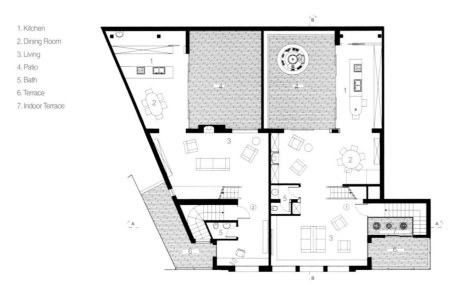

First floor plan

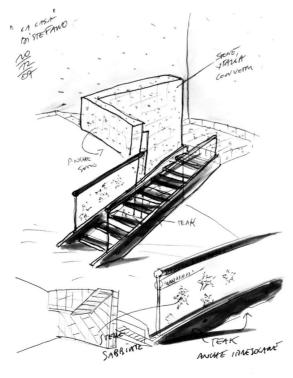

Sketch

H形住宅 H House

Location Maastricht, Netherlands
Site area 300㎡
Building area 130㎡
Total floor area 230㎡
Architecture design Wiel Arets
Architects
Design participation Wiel Arets, Satoru
Umehara, Harold Herman, Daniel
Meier, Dennis Villanueva
Callaborators Alex Kunnen
Francois Steul (Models)
Photography Joao Morgado, Jan Bitter

Designed for a couple and sited in a hilly suburban area of Maastricht near the Netherlands-Belgium border, the H House was designed for a couple with a strong interest in the arts. The clients formerly occupied a home directly adjacent to the site before appointing Wiel Arets Architects to design what would become their new home, the H House. Individually an actor and a dancer, and dually landscape architects, the owners are able to keep their landscaping skills honed in the formal garden behind the house, which they occasionally open to the public.

The interior of the home is comparable to that of a loft-like space with a central mezzanine. Adjacent to the main volume of

the home, are two independent volumes - the entrance and the bathroom, the latter of which is cantilevered over the ground floor. Terraces define the position and shape of the house, each with its own distinct character. The interior of the house has no structural walls and very few rectangular columns support the structural slabs, each positioned to minimize their impact on the interior space. All other walls, whether internal or external, consist of glass in varying shades of opacity. All lighting and bathroom fixtures are parts of the ALESSI Il Bagno dOt series, which is also designed by Wiel Arets Architects.

The stair of the house was conceived as an independent sculptural object within this loft-like setting, producing an

@Jao Morgado

air of suspense, while simultaneously providing storage. The range of privacy desired by the owners resulted in a series of curtains that can be drawn to casually define "interior rooms" on both levels of the house. The combination of transparent and opaque glass, as well as the sliding and fixed portions of the façade, create a number of different possible responses to the changing of seasons and patterns of daylight. The house is suffused with richness due to the layering of unadorned space, the transparencies of the material palette and the countless possibilities of spatial compilations.

@Jan Bitter

@Jan Bitter

@Jao Morgado

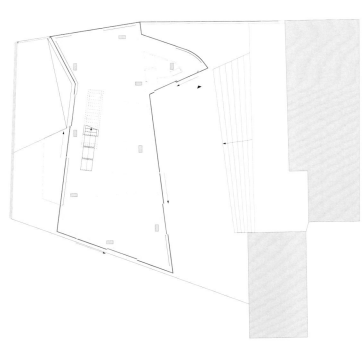

Ground floor plan

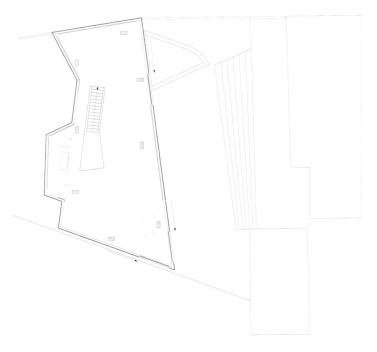

First floor plan

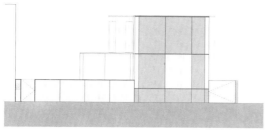

North facade

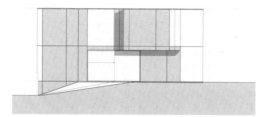

West facade

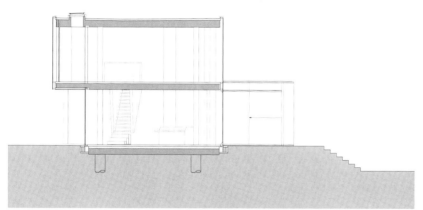

Section A-A

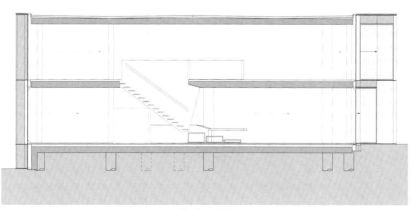

Section B-B

@Jao Morgado

@Jao Morgado

美洲 AMERICA

混凝土住宅 Concrete House

Location Buenos Aires, Argentina
Surface area 502㎡
Architecture design Vanguarda Architects
Photographer Luis Abregú

The main feeling that the home conveys is a sense of defiance of gravity and space. With its large spans of glass that take advantage of the house's relationship to the environment, the space seems almost larger than life.

With an expansive surface area of 500m^2 distributed across 9 rooms, the ground floor is home to the social sector of the house. Here you will find the living room, dining room, kitchen with breakfast bar, and TV room, as well the outdoor barbecue and patio (gallery).

Each space in the house features a special design selected according to the way of life of its owners, and remains true to the rational current style that is perceived in each and every corner.

On the upper floor, find the master suite with a spacious bathroom and private dressing room. At the other end of the floor, two secondary suites are connected by a glass corridor that opens onto a balcony, generating a sense of integration and flow.

The whole house is designed in perfect harmony, with materials such as wood, steel, and glass found throughout. The seriousness of the style is broken by the use of bright colors on some of the walls, as well as on the furnishings.

The union of the spaces creates a sense of endlessness, taking full advantage of the environment, and generates an endless spatiality, which was thought to take advantage of the environment, that ends with an dematerialized surround that covers the architectural box.

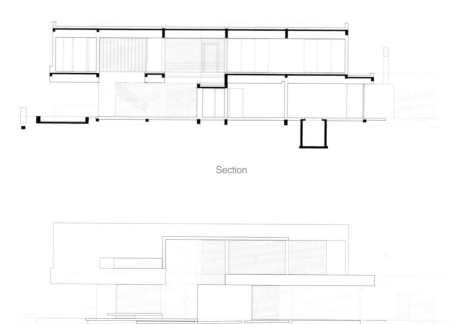

Section

Front elevation

First floor plan

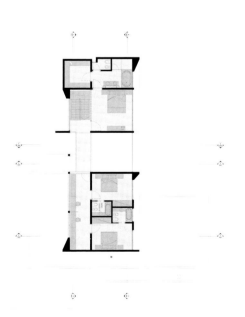

Second floor plan

Q形住宅 Casa Q

Location Misterio beach, Peru
Site area 420㎡
Built area 404㎡
Architecture design Longhi
Architects–
Luis Longhi
Photographer Juan Solano,
CHOlon Photographya
Design participation Carla Tamariz,
Christian Bottger
Construction Longhi Architects –
Hector Suasnabar

Floating in the desert

Infinite rolling dunes from the desert to the East and rocky Pacific Ocean cliffs used by fishermen to the West converge on the site of Casa Q; creating a unique natural environment. Casa Q is the first residence built in one of the areas not yet occupied at the Beach Club Misterio located 117 kilometers south of Lima, Peru.

Challenging the stillness of the surroundings, Casa Q materializes the dreams of a young couple in a "floating volume" which embodies the spaces for a future family. The volume is supported by circular columns placed by intuition, as a dance, instead of forcibly in a grid. The dancing columns are accompanied by sliding glass panels that define the common area of the house; living-dining and terrace are integrated or separated by the option to open or close the glass panels depending on social and weather conditions.

The rest of the house — guest rooms, kitchen and services — are tastefully secluded at the back of the sloped site thus providing visual contrast with a volume of water in the front which has been unearthed for the enjoyment of swimming.

Each view in the house connects to the infinite of the horizon.

1. Bedroom
2. Bathroom
3. Kitchen
4. Dining Room
5. Room

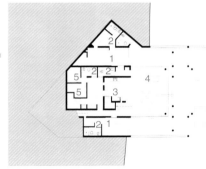

First floor plan

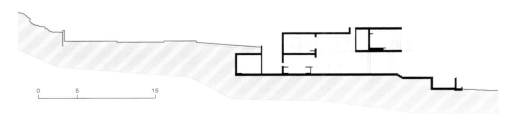

Section

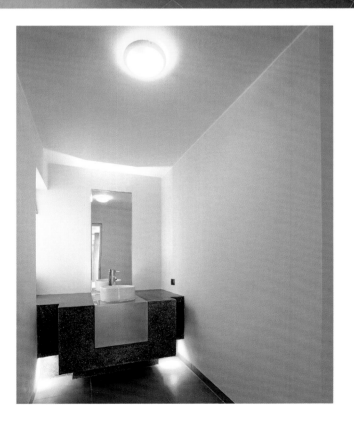

Kona住宅 **Kona Residence**

Location Kona, Hawaii
Size 7,800㎡
Architecture design Belzberg Architects,
Hagy Belzberg, principal Barry Gartin,
project manager
Design participation David Cheung,
Barry Gartin, Cory Taylor,
Andrew Atwood, Chris Arntzen,
Brock DeSmit, Dan Rentsch,
Lauren Zuzack,
Justin Brechtel, Phillip Lee,
Aaron Leppanen
Landscape design Joe Roderick
Hawaiian Landscapes, Inc.
Interiors design MLK Studio
Photography Benny Chan (Fotoworks)

Nestled between cooled lava flows, the Kona residence situates its axis not with the linearity of the property, but rather with the axiality of predominant views available to the site. Within the dichotomy of natural elements and a geometric hardscape, the residence integrates both the surrounding views of volcanic mountain ranges to the east and ocean horizons westward.

The program is arranged as a series of pods distributed throughout the property, each having its own unique features and view opportunities. The pods are programmatically assigned as two sleeping pods with common areas, media room, master suite and main living space. An exterior gallery corridor becomes the organizational and focal feature for the

entire house, connecting the two pods along a central axis.

To help maintain the environmental sensitivity of the house, two separate arrays of roof mounted photovoltaic panels offset the residence energy usage while the choice of darker lava stone help heat the pool water via solar radiation. Rain water collection and redirection to three drywells that replenish the aquifer are implemented throughout the property. Reclaimed teak timber from old barns and train tracks are recycled for the exterior of the home. Coupled with stacked and cut lava rock, the two materials form a historically driven medium embedded in Hawaiian tradition. Local basket weaving culture was the inspiration for the entry pavilion which reenacts the traditional

gift upon arrival ceremony. Various digitally sculpted wood ceilings and screens, throughout the house, continue the abstract approach to traditional Hawaiian wood carving further infusing traditional elements into the contemporary arrangement.

1. Kid's Room
2. Kid's Common
3. Motor Court
4. Guest Room
5. Garage
6. Entry Pavilion
7. Outdoor Gallery
8. Theater
9. Kitchen
10. Great Room
11. Dining Room
12. Office
13. Office
14. Gym
15. Master Bedroom
16. Hot Tub
17. Pool
18. Reflecting Pool

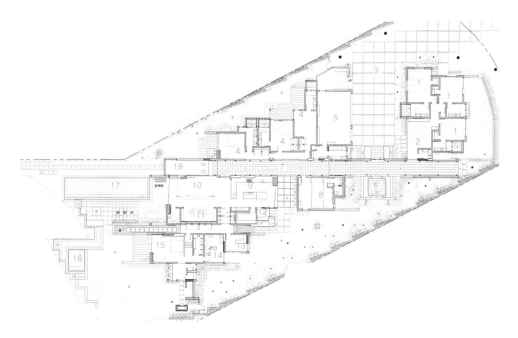

First floor plan

Palabritas海岸住宅 House in Palabritas Beach

Location Lima, Peru
Gross area 230㎡
Architecture design Arch. Jose
Orrego, METROPOLIS
Design participation Arch. Anahi
Bastian
Photographer Elsa Ramirez
Client Mr. Guillermo Santibañez
Planning supervisor Mr. Cesar
Roncagliolo
Lighting consultant Murano light
Main contractor MyM contructores

A summer celebration

The house is located on the first row of lots and was designed so that the architecture acted like a frame to the view of the beach and the islands that are found in front of the house.

The esthetic of the house celebrates the summer with modern curves that remind us of the Brazilian architecture of the sixties.

The whole house has a white finish with color accents in red, in both main elements such as the swimming pool as well as furnishings. The house reminds us of competitive elements used in Stanley Kubrick films, with the white curved elements contrasted with details in red.

The exterior was designed as a white elevated box, exposed from its front side. One of the sides has a concrete lattice based on a contemporary composition with perforations that allow the interior to have a transparency without losing its intimacy.

The dining and living room are placed so that they can be integrate with the terrace by sliding the glass doors.

An interior patio was developed on a lower level where the bedrooms and family room conjoin so that an intimate zone for these rooms were created. The main bedroom is located on the first level and was designed so that it had a view towards the ocean.

As a whole the house created a contrast of red color over white surfaces, in a way that the spaces give an appearance of amplitude.

From the inside of the living room, the architecture offers a frame that shows the landscape of the beach, the island and the sunset.

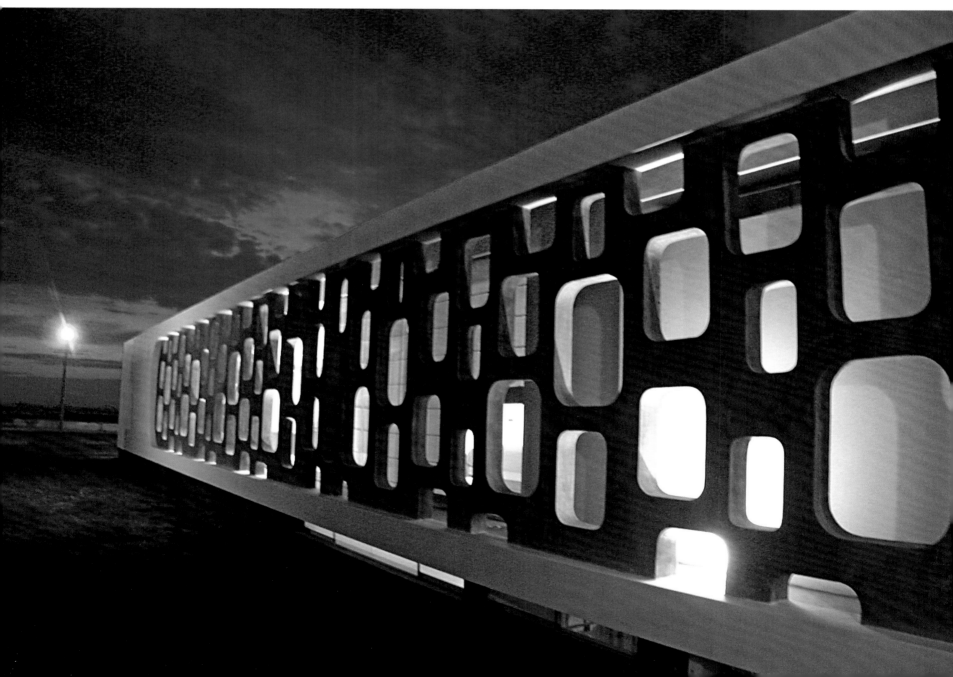

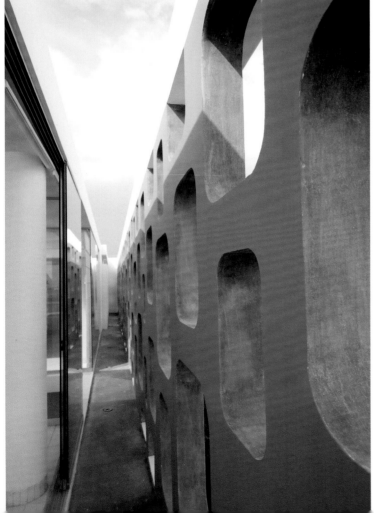

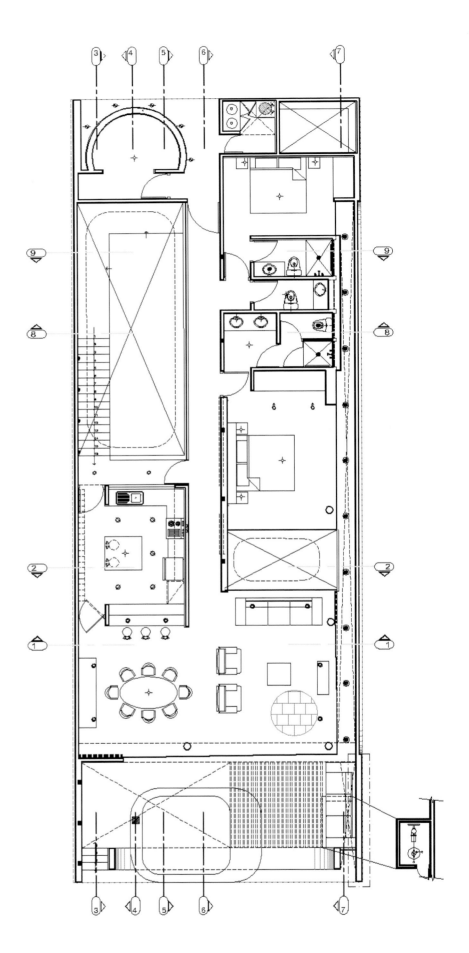

First floor plan

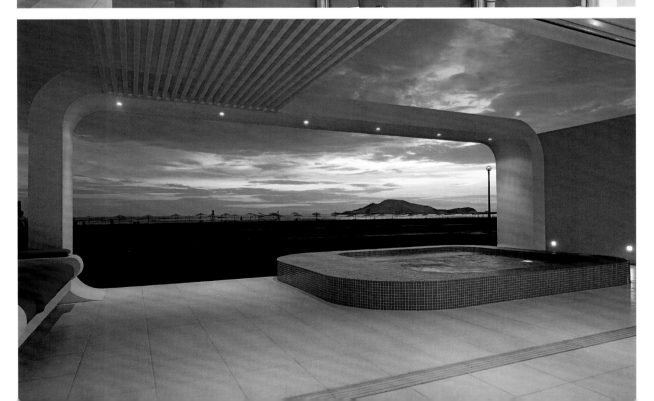

Side elevation

NPT ±0.00m

NTT +3.10m

折叠式住宅 Fold Up House

Architect dba (Dan Brunn)
Shape MERGEFORMAT design branding
architecture

For the client's of this suburban home, the feeling of openness and of continuum was of their utmost importance. As a means to negate the claustrophobic spaces of their primary residence, the design called for a language to express the free flowing nature of the design.

The home is constructed out of folded concrete planes, which in harmony create the illusion of vast open space while still having an intimate relationship to the exterior gardens. This is achieved through descriptive folding and specific openings into the gardens.

Following organic shapes, such as a budding flower, the spaces grow from one another, with open windows on each end of the tube-like extrusions. Within each of the petals the materials are all homogenous, white epoxy flooring and white concrete walls and ceilings. A blurring between what is a wall or what is a floor occurs throughout the home. In the living room, the floor bends upwards to form the lounge area, with its built in cushions for relaxing, then the floor continues to flex upwards to form the walls and the ceiling that create the space. Similarly, this folding occurs in the entry way and the kitchen.

Organizationally, the house is separated into two levels, with the master bedroom at the second floor, overlooking onto the pool area. A folded plane tube stairwell leads one into the master bedroom. Above the stairwell, is the light well, formed by the folding of the stairwell tunnel.

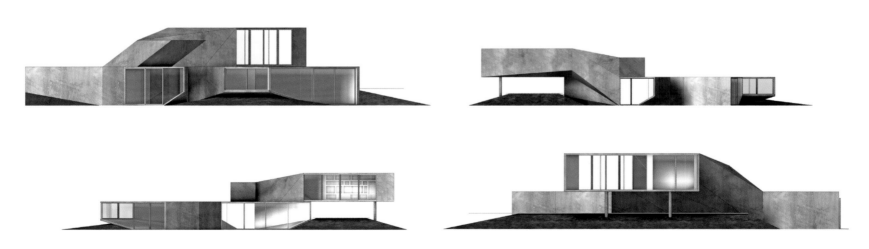

Elevation

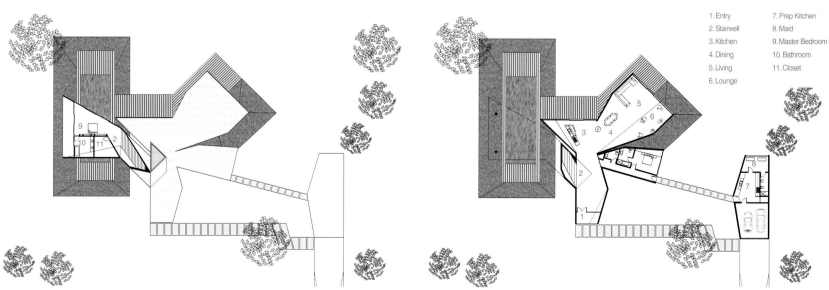

1. Entry	7. Prep Kitchen
2. Stairwell	8. Maid
3. Kitchen	9. Master Bedroom
4. Dining	10. Bathroom
5. Living	11. Closet
6. Lounge	

Site & Ground floor plan

La Encantada住宅 House in La Encantada

Location Lima, Peru
Architecture design ARCHITECT
JAVIER ARTADI
Photographer Elsa Ramirez

Single-family Townhouse at La Encantada
This Project involves a single-family townhouse at La Encantada, a suburb on the Lima shoreline.
Given its sub-urban nature, the house design is nourished with the purpose of seeking the strongest relationship between the property's constructed space and its indoor garden.
For that purpose, a compact space design was thought to contain the program (social area, private area, general service area, etc.) which is intersected by a folded plane.
This architectural folding is intended to connect the house's front garden with the indoor spaces through a spatial round-journey of single and double heights.
In this way, the simplicity of the major volume space is altered, but at the same time enriched with, spaces connecting it with; and freeing it from; the green area surrounding it.

Project description
The project, placed on a corner, is set within a two-storey volume.
The bureau, social area (living-dining room), kitchen and services area are found on the first floor, The stairs, placed in a double-height space, can be seen from the main entrance, its leads to the second floor where the house's more private areas are found; that is, main bedroom, 3 bedrooms for children and a family room. These spaces are connected to an exit to a terrace with a view to the back garden.

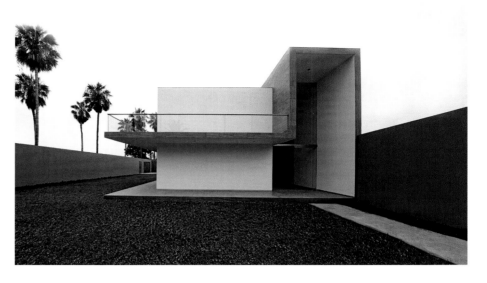

1. Room
2. Bathroom
3. Kitchen & Dining Room
4. Living
5. Study Room
6. Parking

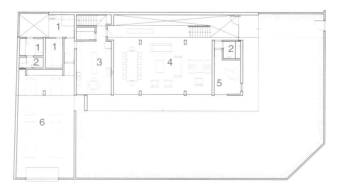

First floor plan

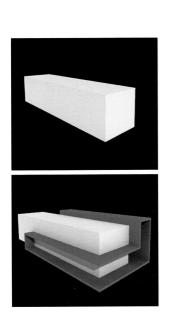

拉斯帕尔米拉斯海岸住宅 Beach House in Las Palmeras

Location Lima, Peru
Gross area 348㎡
Architecture design ARCHITECT
JAVIER ARTADI
Photographer Elsa Ramirez

This project consist of a weekend house located on a beach 100km, south of Lima, the capital city of Peru.

The beach place is a portion of the almost 2,000 km. length of Peruvian coast-line, a very peculiar desert with no extreme temperatures and almost no showers.

This lack of rains has made it possible for local pre-Columbia constructions of sun-dried mud-made bricks with external smooth finishing lasting for centuries.

This so-called "liquid architecture" is also present in modern times in the use of concrete and cement, by obtaining results similar to those obtained with mud-made bricks; that is, an expression of pure and continuous matter and surface.

As in all my previous projects rendered along the Peruvian coast, have sought this characteristic to be reflected in the most conspicuous portion of the design of the house.

Conceptually, the house is kind of a tray in white which seems to be suspended on a stone pedestal.

The materiality of this major feature is an object fully reflecting smoothness and purity resulting from fine finishing in hand-made cement covering.
The internal distribution of the house is matched with the volumes comprising it: a first level used for service areas and

garage; a second level (acting as a hinge) where bedrooms are found; and a third level intended for the social areas: living room, dining room, terrace and swimming pool.

The project is also intended to establish a very direct relation with the nature, with earth, heaven and sea. Also, while a person goes up through the interior, transparent staircase allows inhabitants to view the backyard hill, made of stone and sand, and one on the third level,. a carefully studied terrace allows for the connection between heaven and the sea horizon.

Ultimately, the house reminds us of the human presence within a natural environment in the form of a simple and primary geometry that contrast with the beautiful landscape of the desert Peruvian coast-line.

1. Bathroom
2. Guest Room
3. Laundry
4. Storage
5. Storage
6. Garage
7. Entrance

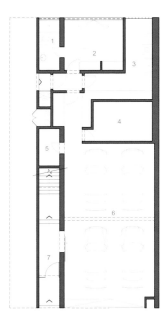

First floor plan

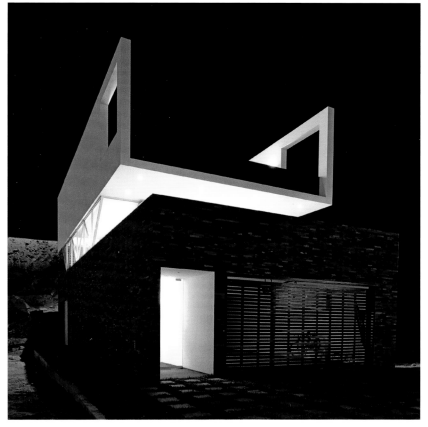

H2o滑雪住宅 Ski House H2o

Location Florida, USA
Size 3,100 m^2
Architecture design hughesumbanhowar architects
Photographer Ken Hayden, Tom Winters

The client is a water ski enthusiast. Banding together with several fellow enthusiasts, they purchased some agricultural land and modified an existing artificial lake to be used for waterskiing/slalom course racing. The house sits in and on a berm dredged from the bottom; the design hence engaging the lake in multiple ways. This elevated perspective in an otherwise flat surrounding topography provides a view of the 1/2 mile watercourse from end to end.

The house is a simple bar conceptually and literally broken into 3 use. defined areas; i.e. guest wing, public/fraternal spaces, master bedroom. A wing-shaped roof hovers above these 3 parts to encourage circulation movement from one space to another. The roof design also provides shade to the lakeside terrace, while simultaneously capturing and directing the water-cooled breezes through the upper transom windows of the living space. This "free" and passive conditioning reduces the use of mechanically driven conditioning. Rain collection from the valley of the roof is directed to an underground reservoir to be used for irrigating the native planted landscape.

Materially modest components such as corrugated roof SIP panels, concrete floor, Stucco exterior or exposed interior concrete block walls, aluminum clad windows, cloth roll down shades make for simple and lean construction: a suitable response to the client's request for a dramatic, humble, inexpensive, low maintenance, yet inspired shelter.

© Tom Winters

© Tom Winters

© Ken Hayden

© Ken Hayden

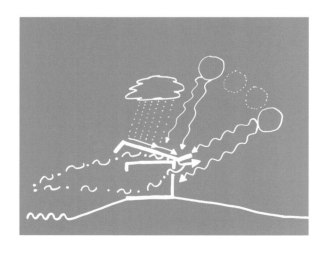

Sketch diagram

North elevation

© Ken Hayden

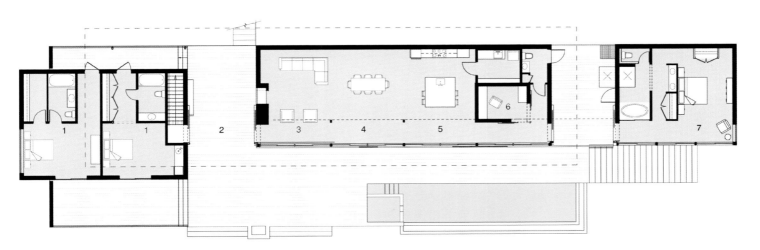

1. Bedroom

2. Entry/Breezeway

3. Living Room

4. Dining Room

5. Kitchen

6. Office

7. Master Bedroom

Floor plan

圣神阿鲁马住宅 Santo Amaro House

Location Sao Paulo, Brazil
Site area 4.095,51㎡
Built area 1.274,86㎡
Architecture design Isay Weinfeld
Collaborator architect Domingos
Pascali
Project manager Monica Cappa
Santoni
Design participation Flavia Oide |
Gustavo | Benthien | Leandro |
Garcia Alexandre | Nobre
Photographer ©2011nk@
nelsonkon.com.br

Santo Amaro House was designed for a mature couple with grown-up children that no longer live in the house. It was built in quiet residential area in Sao Paulo, Brazil.

The homemaker is an amateur piano player who enjoys gathering friends at home for intimate musical recitals. For that purpose, a wide room with high ceilings was designed to accommodate three Steinway pianos, including a concert grand piano, spacious enough for small musical ensembles, such as, for instance, duos, trios or quartets. That environment also features a winter garden receiving natural lighting through a glass roof.

The living room and dining room are separated by a low screen, and a sliding door separates the living area from the den, thus allowing the husband to read or watch a film without disturbing his wife during piano lessons.

A deck that externally surrounds the living and dining rooms and extends along the swimming pool can also be reached directly from the main bedroom.

Private quarters feature also two spacious guest suites, in order to accomodate the couple's daughter and family, who live abroad but come visit often.

Service areas - garage, deposits, employees' quarters, laundry, etc - were all set up on the ground floor.

The existence of three fully-formed brazilian ebony trees defined the set-up of the building, as to preserve the trees: the dining room overlooks an atrium built around one of the trees; the volume of the house and the main deck, had their limits defined by the remaining two brazilian ebony trees.

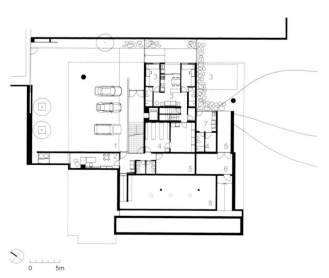

Ground floor plan

0   5m

1. Garage
2. Office
3. Servants' Living Quarters
4. Deposit
5. Mechanical Room
6. Workroom
7. Laundry
8. Salon
9. Pianos' Salon
10. Garden
11. Sitting / Dining Room
12. Family Room
13. Deck
14. Private Living Room
15. Bedroom
16. Closet
17. Kitchen
18. Swimming Pool
19. Sauna

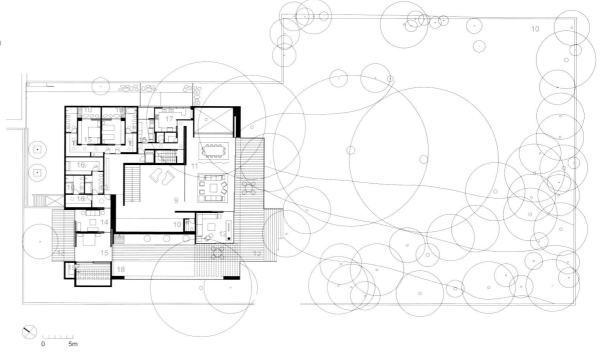

0   5m

Upper floor plan

树林住宅 Phinney Residence

Location Seattle, USA
Area 233 m^2
Architecture design
Elemental Design, LLC
Contractor Logan's
Hammer
Structural LFD Structural
Engineering

The clients, a young family of two adults and one child desired a modern home, much like the ones they fell in love with on their frequent trips to Scandinavia. After visiting a number of different sites, they settled on one in the cozy Seattle neighborhood of Phinney Ridge. Known for its classic Seattle craftsman homes and beautiful views of the Ballard Locks and the Olympic Mountains, the selected site was ripe to make a bold design statement.

It was, in fact, the lots irregular footprint that shaped the design of the residence. With a north to south slope and an existing driveway that cut through the lot, only 117m^2 of the lots 467m^2 was deemed buildable. It was known early on in the design process that the slope of the site encouraged a three-story structure. Add in the fact that based on Seattle Building Code, the minimum height over an existing driveway can be no less than 16 feet. It was this restriction that led to the third floor cantilever, which, not coincidently is the residence's most striking feature.

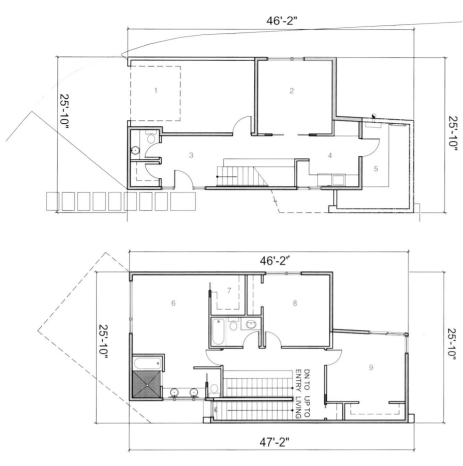

46'-2"

25'-10"

25'-10"

1   2

3   4   5

46'-2"

25'-10"

25'-10"

6   7   8

9

DN TO ENTRY   UP TO LIVING

47'-2"

Ground & First floor plan

1. Garage
2. Studio/Office
3. Entry
4. Laundry
5. Mechanical/Storage
6. Master Bedroom
7. Master Closet
8. Bedroom 2
9. Bedroom 3

1. View Deck
2. Living
3. Dining Room
4. Pntry
5. Family Room
6. View Deck
7. Green Grid Over
   TPO Membrane
8. Roof Deck

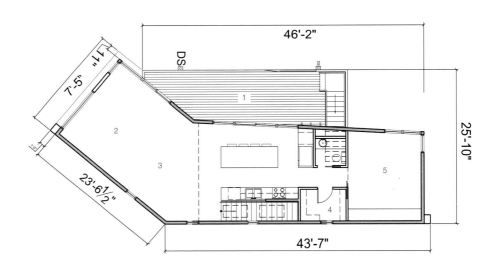

46'-2"

7'-5"

1'-1"

23'-6 1/2"

25'-10"

43'-7"

Second floor plan

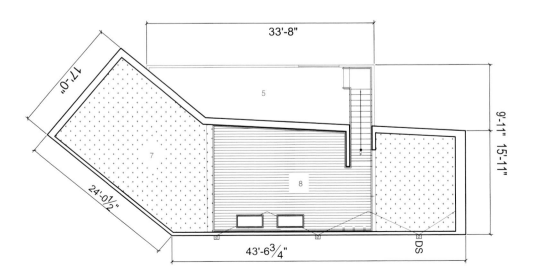

Roof plan

Mikve Rajel住宅 **Mikve Rajel**

Location Lomas de Tecamachalco,
Cuidad de México, México
Interior Design Pascal Arquitectos,
Carlos Pascal and Gerard Pascal

Construction Rafael Salame

Furniture Pascal Arquitectos

Photographer Víctor Benítez

The Mikve is the ritual bath of purification in the Jewish religion. It is possible diving in fresh spring water, or in a place specially dedicated to it, fed by rainwater that must be collected, stored and communicated to the vessel that is called a Mikve. All this must be made under a very strict set of rules related to the degree of purity of water. These rules also include the use of materials, architectural measures and water treatment.

The Mikveh is mostly used by women once a month, and for the brides to be, for conversions and certain holidays.

There is also a Mikveh used for the purification of all elements of kitchen and food preparation.

Mikveh is known to represent the womb, so when a person enters the pool, it's like to return to it, and when it emerges, as if

reborn. In this way, you get a totally new and purified condition. Its symbolism represents at the same time, a tomb, therefore, can not be performed the ritual bath in a tub, but must be built directly into the ground. The fact that illustrates the Mikveh much as the woman's womb and at the same time as the grave, becomes not a contradiction, since both are places where you can breathe, and at the same time are endpoints of the cycle of life.

This project has a special meaning for us. 20 years ago we designed Mikve. Rajel, it was the first "designed" Mikve, there were other such places but not very fortunate, dirty and neglected, community members were not going any longer, and the ritual was dissapearing, which according to the Jewish

religion is the most important.

When designing the mikvah Rachel we really did not know which would be the consequences of its actual use, how the event was to going to unfold. It was so successful that all the communities began to make their own Mikves, but more than mystical spaces they seemed like luxury spas.

Today 20 years later we realize that the event for the brides becomes a big celebration, and that there was not room for all the assistance, plus 20 years of use is also influenced by architectural trends of the moment puts it out of time and so we have to destroy it to create a new proposal that meets the changing needs , both aesthetic and use.

The reception becomes a big box of white light, suggesting purity., No columns just delicate natural aluminum vertical beams and white glass, floors in Santo Tomas marble, modern and sleek white sofas and starring up the wall and turning on the ceiling, a mural of the artist Saul Kaminer.

The corridors around the building giving access to washing bathrooms and from these in to the Mikve, it must be a separate access and exit, as you enter impure and exit pure, , contrasting the dark on the floor and walls and ceilings in white enhancing the visual drama with recesed lighting .

The bathrooms are marble-lined in Santo. Thomas unpolished and polished white glass, stainless steel furniture and Arabescato marble, from here we enter to the Mikve, tall space with a gable roof of which is collected water to be used in the ritual. Cumaru wood paneling, marble floors and St. Thomas lined pool.

1. Bathroom
2. Makve
3. Toilet
4. Bridal Room
5. Garden

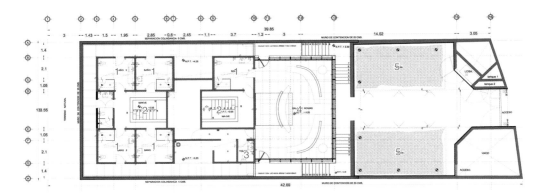

Floor plan

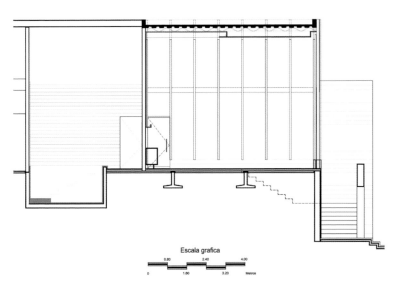

Section

德勒住宅 Zeidler Residence

Location California, USA
Site area 768㎡
Total floor area 450㎡
Architecture design Ehrlich
Architects
Interior design Ehrlich Architects
Photographer Matthew Millman

Located on the bluffs overlooking the Pacific Ocean, the 450㎡ Zeidler Residence arranges interior and exterior living spaces to maximize views, natural light, and ocean breezes within a subtle, sophisticated material palette.

Designed for a retired couple with grown children, the house sits on a relatively flat corner lot with expansive views of the Pacific and vegetated cliff that leads from the site, down to the beach. The parti divides the program into two main structures connected by a sheltered courtyard. On the ocean side, the two-and-a-half-story main house features a double-height living space, with full-height glass doors that open the interior onto the exterior spaces. A mezzanine is oriented towards the view. At the end of the stair tower, a full-sized roof

deck accommodates various entertaining configurations and provides strong connections to the landscape and views beyond. The front yard even incorporates a petanque court, a favorite pastime of the client.

The rear structure accommodates separate living quarters for friends and family in three oversized bedrooms. The second level studio has a full kitchen and expansive deck with views towards the ocean. The two primary structures frame a landscaped courtyard with lap pool and built-in barbecue, and when opened to the elements form a complex of open air pavilions connected through the landscape.

1. Entry
2. Family Room
3. Outdoor patio
4. Dining Room
5. Kitchen
6. Pantry
7. Pool
8. Guest Room
9. Garage
10. Petanque Court

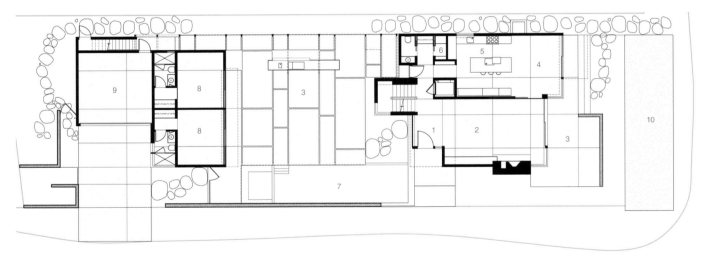

First floor plan

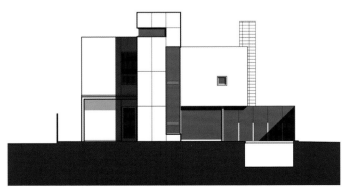

Northeast elevation

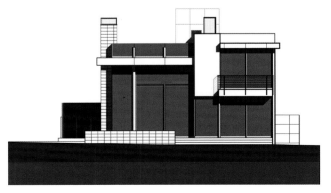

Southwest elevation

秘密宾馆 Secret Guest House

Location Chimalistac, Mexico
Site area 1,249.90㎡
Floor area 624.42㎡
Architecture design
Pascal Arquitectos
Designer Carlos and Gerard Pascal
Interiors, Landscape, Lighting,
MobilAire & Construction Pascal
Arquitectos
Photography Pascal Arquitectos

Contemporary architectural family house in a residential classified historic colonial zone of Chimalistac in Mexico city. It difficult to say anything else about another house, and this is what this is about, is a house made to order for a client, which is often more complex than developing a large building. The result depends on two factors: a good architect, but rather more of a good client.

The context in which the house is inserted has an historic colonial character untouched by the unorganized sprawl that has occurred elsewhere.The fact of intervening in a historic area entries (INAH) raises the dilemma of adapting or blending into the context , but by ideology could not talk about the present and future with the language of the past. Fact generated some discussion with the INAH (National Institute of Anthropology and History) http://www.inah.gob.mx/

because their vision is to recreate the past even if this results in a pastiche , and finally and where the interaction between the inside and outside occurs our goal was to achieve the neutrality that would make the transition from the historical to the modernity of the interior.

The fashionable and politically correct slogan nowadays is that everything must be sustainable over yet despite the good wishes and intentions, and after several runs to determine the relationships financial cost benefit, the only we could manage to improve was in energy saving light bulbs and intelligent control systems and sensors linked to timerand more efficient irrigation systems.

The most important was the use of intelligent design to make the house better in comfort and climate, and the building

design process in which no processing or transformation of materials such as stone, wood etc.was done, a system based on just in time logistics and a change in how the jobsite is managed ,with prefabrication and the inclusion of pre-finished items.

As parts of the architectonic discourse and for reasons of durability and maintenance very few finishings were used, being concrete one of the most notable of the house for its ability that does not age and decay, and the fact it acquires more dignity and history with time.

One of the main objectives was to achieve the most natural light and views to the garden, and not to create a series of closed rooms but a series of spaces where the events happen and articulate wit one another. It is important to note that the entire house is designed in modules and multiples of feet and then generating different size of overlapping rectangles, that became the generating pattern of the geometric theme of the house.

Site plan

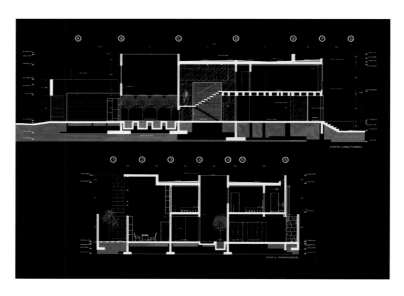

Transversal section

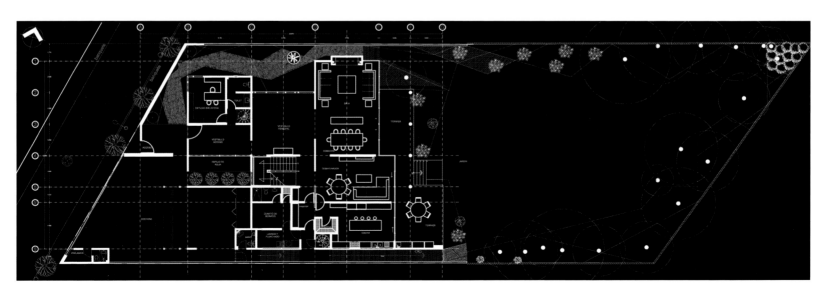

Site & Floor plan

卡拉拉住宅 Carrara House

Location Buenos Aires, Argentina
Area 600㎡
Architecture design Andrés Remy Arquitectos
Design participation Andrés Remy, Hernán Pardillos, Lilian Kandus, Diego Siddi, Gisela Colombo.
Landscape design Leandro La Bella,
María Celeste Iglesias
Lighting design Mauricio Meta
Engineer Carlos Dolhare
Photographer Alejandro Peral

Located on an irregular lot, the house sits at the back of the lot and is parallel to one of the streets to open the best orientation and capture the best views. The idea of this journey was to discover the entrance as we follow the exterior stone wall. The rustic and crafted stone defines and separates the entry zones from the living spaces and is inside and outside, proposing a counterpoint to the pure white that dominates the inside of the house.

A blind and evocative entrance makes a strong impression on the house. The white of the carrara marble dominates the interior architectture. With the white walls and cielings, the house appears to arise from within the water. The touches of color are used for small details and decorative objects, dominating the white color and the turquoise of the water.

The water that surrounds the house penetrates it in the form of the mirrored surface of the water whose novelty results in the interior cascade that emerges from the top floor and falls while painting reflections via a pane of glass. This mirror of water is reproduced outside blurring the boundaries between one and the other. Finally, emerging from the hall dispenser of the top floor, the glass cascade drains musically into and through the heart of the ground floor. These elements give the project the distinct mark of Remy-bold, creative and perhaps provocative, but always unique.

1. Entrance Hall
2. Living Room
3. Dining Room
4. Kitchen
5. Guest Bathroom
6. Laundry
7. Service Bedroom
8. Expansion Room
9. Barbecue Room
10. Swimming Pool
11. Water Mirror
12. Master Bedroom
13. Master Bathroom
14. Bedroom
15. Bathroom
16. Studio

First floor plan

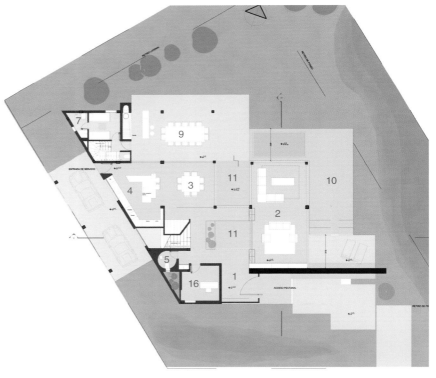

Ground floor plan

PLANTA BAJA   PRIMER PISO

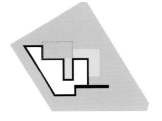

Natalia住宅 **Casa Natalia**

Location Guadalajara, Jalisco, Mexico
Site area 642㎡
Construction area 565㎡
Architecture design Agraz Arquitectos
Designer Ricardo Agraz
Photographer Mito Covarrubias
Collaborators Erick Martinez, Jessica
Magaña, Javier Gutiérrezavier Aguirre,
Israel Picos, Juan Antonio Jaime
Art Adrian Guerrero
Industrial design Cubo 3, Hector Navarro

It can be said that Casa Natalia is a briefing of the Agraz Arquitectos policies since it gathers the complying conditions for the firm's main features. In an outstandingly oriented North-South terrain, a single longitudinal volume was designed, adding a limb to stabilize it. Then, as done before, the program begins by taking the cars out of the architectural scenery and placing them underground, which optimizes the surface area of the land piece.

And from this perspective, we repeat the placing of the basement a half level lower and lifting the house another half level, upcoming to a garage that shares the houseplant with service dormitory, laundry and equipment areas. From this point, a stairway goes upwards communicating all three stories of the house, with the same finishing of the rest of the areas, immediately leaving the garage environment.

As in other programs of the firm, the first flight of the stairway leads to the main door, which by being separated from the car entrance, leaves a front plaza for the house that dilutes all frontiers between urban and architectonic spaces.

Once inside the first floor, the living and dinning rooms offer an extension with an intimate family room and a terrace that can be the perfect social place due to its transforming possibilities: it can be fused or isolated from the rest of the precincts according to the needs and has an independent entrance.

This time, the kitchen becomes the gravity center of the project giving service to the dinning room as well as to the terrace, whereas the guest bathroom is located in a fair distance to give a comfortable privacy to its user. As a compliment of this houseplant, there is a guest room that is contemplated for the probable future dwelling of the house owners.

Configured for a family made up of the parents and an only daughter, the upper floor is in this sense different from others. There are only two rooms with atypical dimensions as for its spatial generosity, and a reading room, gymnasium and storage room that make up the most out of this small length terrain according to the client's needs.

The vertical circulations that join all stories are contained in this added limb and coated inside and outside, by metal and wood shutters designed by the artist Adrian Guerrero. These control light and privacy and allow a poetic dialogue between glass and steel.

Natalia House, a single volume with an added rib that ends up being the main figure of the program, a piece of architecture bounded to relate to the everyday desires and traditions of this particular family.

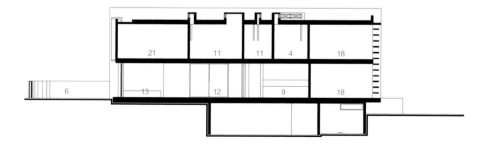

Section B-B

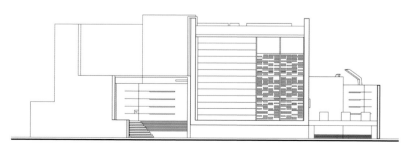

Main facade

1. Garage
2. Service Room
3. Laundry
4. Bathroom
5. Ramp
6. Garden
7. Main Entrance
8. Lobby
9. Kitchen
10. Pantry
11. Closet
12. Dining Room
13. Living Room
14. Terrace
15. Family Room
16. Gym
17. Reading Room
18. Principal Bedroom
19. Service Room
20. Patio
21. Bedroom 01

First floor plan

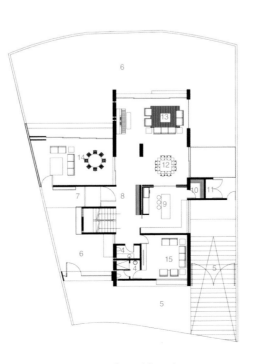

Ground floor plan

S形别墅 House S

Location Rio Negro, Argentina
Site area 2,645㎡
Built area 591㎡
Architecture design Architects
Carlos Galindez, Santiago Alric,
Federico Lloveras
Photographer Sosa-pinilla,
Alricgalindez arquitectos
Staff Alfredo Quiroga, Sofía Peluffo,
Joaquín Adot
Structural engineer Jorge Zapata
Specific advisement Arch.Juan
Carlos Beverati (Sanitary), eng.
Edgardo Gaviño / Arch. Julio Nieto,
(thermodynamic facilities), Arch.
Martin Evans / Arch. Silvia de
Schiller (Bioenvironmental design)
Construction Arrieta & Arrieta
construcciones

The house is a result of several stone boxes piled up on the side of a sloping area. Each one of these cases points and frames an outstanding view, where we can find Catedral, Otto and Ventana mountains, as main characters of the landscape seen from inside.

The space is organized so that it can be divided in different zones, depending on the number of persons that are actually using the place.

We used local stones to keep a coherent language with the harsh and wonderful surrounding, trying to create the idea that the house was always there, just as another rock or an unnoticed natural feature.

The social area is placed on the access level, dominating the best views captured by white calm boxes from inside, opposing to the overwhelming exterior. The master bedroom and the guests room share this level too, each facing a different mountain, in this surprising kaleidoscope of sights.

On the ground floor level, the rooms for the kids, shaped by the land slope, creating different situations and volumes, and a two level loft, due to a sudden depression on the terrain.

As an extension of the deck, a sinuous wooden stair guides us to the last box, resting a few steps below. A sightseeing Jacuzzi offers the best sunset views, and below this, a warm gym and a sauna complete this relax an inspiring lookout.

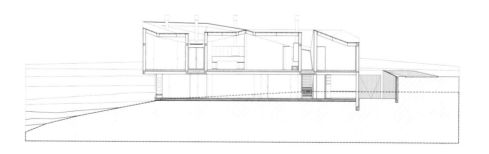

Elevation I

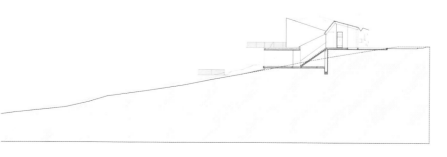

Elevation II

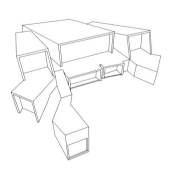

Axonometric

1. Kitchen & Dining
2. Living
3. Bedroom 01
4. Bedroom 02
5. Bathroom

Floor plan

CC别墅 Casa CC

Location Lima, Peru
Site area 383㎡
Built area 389㎡
Architecture design Longhi
Architects – Luis Longhi
Design participation Carla
Tamariz, Christian Bottger
Construction Longhi
Architects – Hector
Suasnabar
Photographer Juan Solano,
CHOlon Photography

Hard and soft connected by light

Two volumes, crown the top of a hill at Playa Misterio, are a gated beach community, 117km south of Lima.

One cube built in exposed concrete comes from the earth while the second one located 2.40m apart and made out of glass as if it is "floating" on the site.

The "concrete block" houses the social area of the dwelling in its lower levels and the master bedroom at the top. The "glass block" accommodates the children's area in two levels.

The space between the blocks is occupied by a staircase designed as a series of sculptural elements that connect the hard appearance of concrete to the soft look of glass, provoking light and shadow effects as the sun visits the house during different hours of the day.

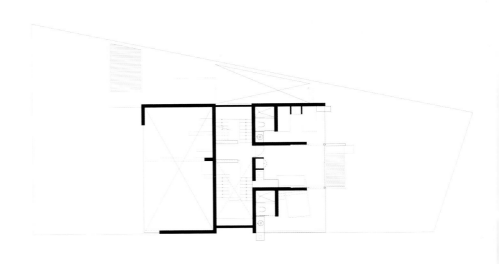

Third floor plan

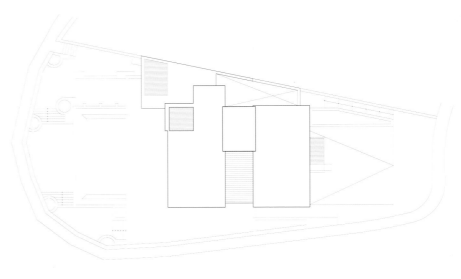

Section

First floor plan

天浪住宅 Skywave House

Location California, USA
Site area 445㎡
Bldg. area 278㎡
Total floor area 210㎡
Architecture design Coscia Day
Architecture and Design / Anthony
Coscia
Design participation Anthony Coscia
project designer, Elita Seow 3D CAD
Interior design Coscia Day
Architecture and Design
Photographer Erhard Pfeiffer

"Anthony Coscia", a design partner and founder in "Coscia Day Architecture and Design", created for himself and his wife Grace Suh a 209m² home which feels more like 334m². The increased sense of space is accomplished with long skylights, large glass walls, an enclosed outdoor living room and the multi-level open plan. Rooms of glass and open spaces flow into one another visually and spatially, while the temperate Southern California climate allows for a further blurring of the interior and exterior space. Sliding glass walls, resin panels and moveable interior partitions can open to reveal even private areas to the rest of the house and nature beyond. A true indoor outdoor idea of living is achieved both at night and day. "Skywave House" is a theoretical work of art. An object occupying the three-dimensional air

space of the site, it has a closed yet unfolding face to the street. The view from the rear North elevation allows one to partially understand the house as a conceptual single plane transformed into a wave that wraps the upper levels. The open-ended reading of the house from the rear connects the interior space to the exterior space above the neighboring rooftops, while leaving the possibility of another segment of the home to be built in the future. The architectonic form of the house utilizes the design language of the Fold that Coscia Day has been developing for almost two decades which has been influenced by two Asian art forms including Origami fold techniques and Sumei. The two dimensional calligraphy concept of Sumie writing a word without taking your pen from the paper, which if

examined in three dimensional space, produces a continuous surface plane of straight and curved segments of differing thickness. Other traditional Asian architectural concepts were incorporated, including the idea of Shoji screen sliding panels as doors or walls to break up a larger space and a raised floor plane elevated above the earth as well as the Korean concept of a raised heated floor. The house is conceptually cut from a single 2 dimensional piece of paper, which is then folded into a 3 dimensional form with distinct sectional qualities. The structural skin wraps up and over itself manifesting into 3 separate roofed volumes, while simultaneously bending into itself to become the floating front living room.

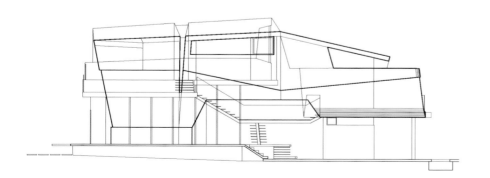

West elevation

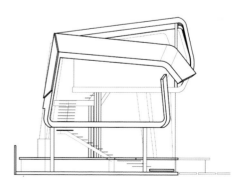

Section

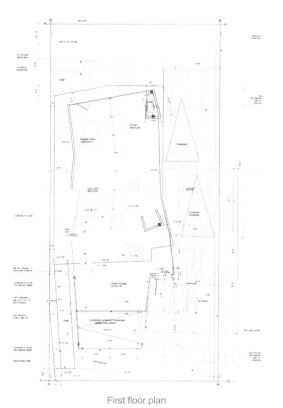

First floor plan

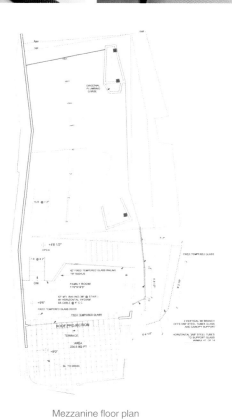

Mezzanine floor plan

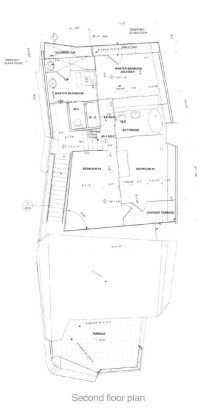

Second floor plan

EL三角形住宅 EL Triangulo

Location Guanacaste, Costa Rica
Project Area 550m^2
Design company Ecostudio Architects
Architecture design Roberto Rivera,
Ana Ulloa
Photographer Anny Leiva

Triangulo [Spanish for Triangle], as the name states, is based on three line geometry. Each line has different length and the two largest sides delimit the property and point to steep slopes and abundant natural richness. The sharpest angle points emphatically to the West coast of Ostional, Guanacaste. The project's shaping consists of a volumetric set that mixes and contrasts filled and empty spaces, horizontal and vertical, marked by a surrounding main solid element that fades away until it visually merges with the natural context. In the same way the entrance is marked by an impressing grey volume with elegance and sobriety which anticipates the access to an open and transparent space. Upon crossing the threshold, spaciousness is expressed by double heights that end on a skylight, which illuminates simply the water running.

Rooms, circulations and nature are joined within a concept of lightness and transparency through the absence of solid elements while the staircase emerges extremely light drawing on a single stroke the merging of the levels. It erects as a sculpture element in the social area.

Conceptual exploration of boosting the relationship between indoors and outdoors generates spaces that completely open to the outside, allowing a perceptual merge of the first level which causes a feeling of being immersed in the canopy.

The architectural proposal seeks for space with surrounding lines together with contrasting angles to break free of the orthogonal, expressing in this way the dynamism of the few walls that make up the residence.

The use of color materials and textures is subject to creating integration and a contrast with the environment at the same time. This ranges from the tonality chosen for the windows to the volumetric language conceptualization and its corresponding characteristics.

Within this conceptual framework, different spaces acquire their own language by themselves keeping general influences of the architectural proposal but with particular variances that enhance their spatial quality. The kitchen joints the social space but it stands out to achieve a particular clean and minimalistic shape. The living room joints it and this continues to a wide

terrace generating an integrated triangulation but with spatially different characters.

Subsequently, after finishing the visual tour in the first level, attention is drawn towards the sea that merges with the swimming pool and its endless edge.

The upper level further explores the relationship with the panoramic view of the ocean, framed by hidden evidence that show the project's constructive lightness. Besides, the dry gardens over flagstones and the upper sun deck stand out as an extension and complement to the interior in order to make a better use of the visuals.

Particular spatial and shape elements come up as complements. For example the runway-deck that floats over the steep slope that goes around the residence. This works as a connecting element for the guest room and the social area, as a link between the outdoors and the indoors. The Gazebo also stands out in function of the swimming pool, and it is perceived as an object that is part of the landscape.

In the same way there is an interest for reducing the impact

with the natural environment aiming to be compatible with it and to be energy efficient. These are the reasons why there is a functional implementation of solar panels for water heating, wide monitors for temperature control that crown spaces with more use, such as the kitchen and the master suite, and they allow extraction of hot air providing cross natural air flow, thus minimizing the use of air conditioning. Same thing happens with the window selection where large sliding panes from floor to ceiling allow for maximum opening.

In this way bio-climate control plays a mayor role in the project, e.g. using generous overhangs to reduce sunlight exposure and work as protection of heavy and constant rains typical in the place.

In summary, triangle moves from the natural to the built; it lies in the peace and harmony of the natural landscape that hosts it while resting as a guest that is respectful and at the same time enhancing and interacting with the exuberant scenic level that surrounds it.

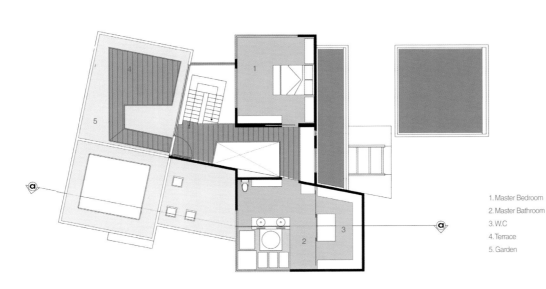

Second floor plan

1. Master Bedroom
2. Master Bathroom
3. W.C
4. Terrace
5. Garden

林中小亭 Pavilion in Wood

Location Mexico
Total area 80㎡
Altitude 1,830m above sea level
Architecture design Parque Humano
Interior design Parque Humano
Landscape design Jerónimo Hagerman
Lighting design Parque Humano
Design participation Jorge Covarrubias + Benjamín González Henze
Photography Paul Rivera, ArchPhoto

Space of retreat and meditation

Pabellon en Lamina is a space which combines the provision of temporary shelter with an inducement to participate in specific acts of memory, contemplation, and philosophical speculation; as well as a place for retreat and meditation. It is related to the process of creating philosophical works, which may take as their subjects – the nature of the environment in which the pavilion itself is sited.

The Pavilion is made in relation to architectural form as well as to the sculptural and visual senses. In its sitting, this pavilion is deliberately positioned in a large plot of land, following the path of an existing pine tree alley.

The view of the inside is in constant flux according to the conditions of changing light and the position of the sun, which affects the reflection and transparency of the glass. Perception has no time spam, there is no acknowledgment of temporality, the art experience is pure. The observing subject is conscious of being part of a present palpability, located in a specific time and reality. For the concept of this project, we sought for a higher unity between architecture and nature.

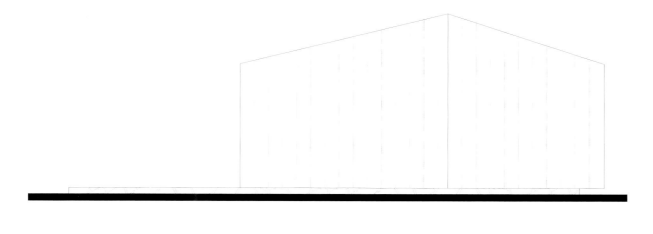

East elevation

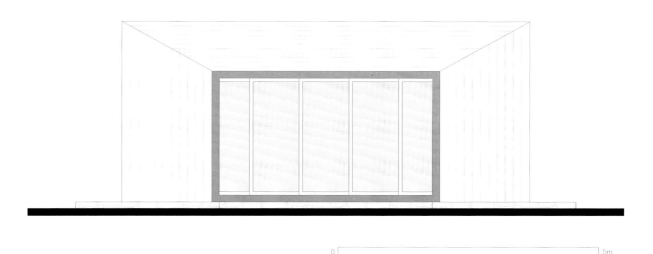

North elevation

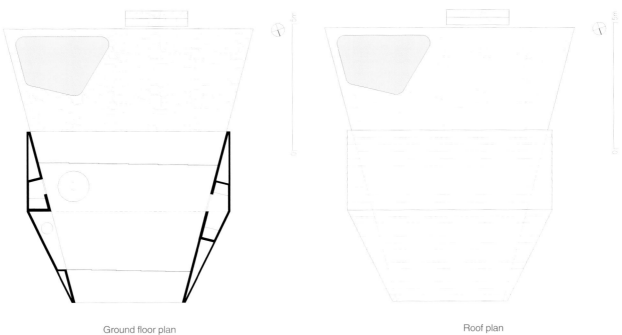

Ground floor plan

Roof plan

带状住宅 Ribbon House

Location Rio Negro, Argentinean
Patagonia
Architecture design G2 Estudio
Photographer Laila Sartoni

The initial idea comes from the juxtaposition of volumes, each containing different functions, on one hand the social life and on the others the private life. When this volumes meet each other, mixing the geometry and the space, it generates dynamic routes between the activity and rest areas of the house, that getting in tension they experiment the transition between being supported on the rock to raise into the sky searching perfect visuals. Is this way that we can appreciate an up-down experience link.

The morphology and materials used, were thought to achieve that the strong became in fragile, the solid in ethereal, the supported in support, the dynamic in static, and vice versa.

So the house is a search between the balance, juxtaposition, ribbon, viewing-point, vital tour, and hug. To get the artistic expression and for reach the limits of the materials, the work was performed with two different systems that can reflect the idea of the project. The support would be reinforced concrete, bringing it to their fullest potential in horizontal planes, vertical and lead off, for a seismically active area such as San Carlos de Bariloche, along with the stone as a heavy and rustic material in dialogue with the nearly mountains

The morphology and materials used, were thought to achieve that the strong became in fragile, the solid in ethereal, the supported in support, the dynamic in static, and vice versa. So the house is a search between the balance, juxtaposition,

ribbon, viewing-point, vital tour, and hug. To get the artistic expression and for reach the limits of the materials, the work was performed with two different systems that can reflect the idea of the project. The support would be reinforced concrete, bringing it to their fullest potential in horizontal planes, vertical and lead off, for a seismically active area such as San Carlos de Bariloche, along with the stone as a heavy and rustic material in dialogue with the nearly mountains.

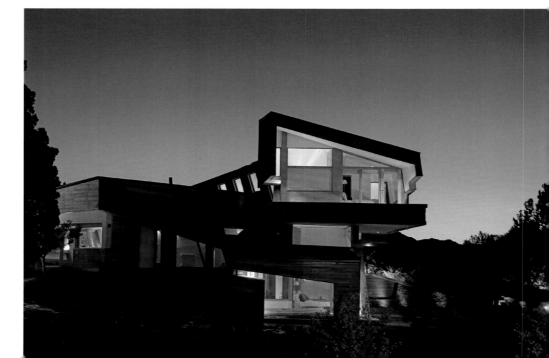

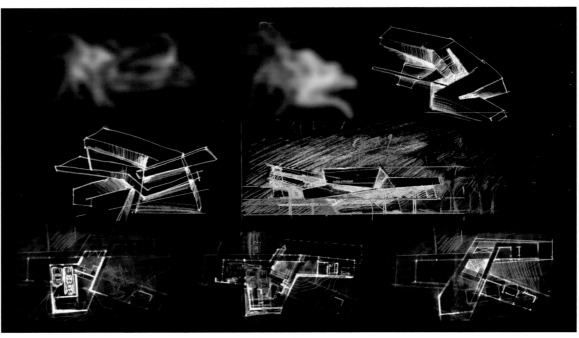

Process Sketch

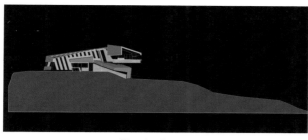

Elevation I

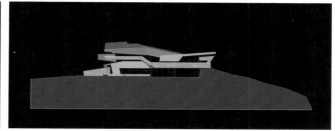

Elevation II

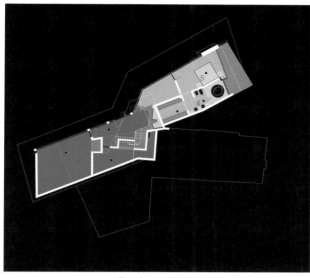

First floor plan

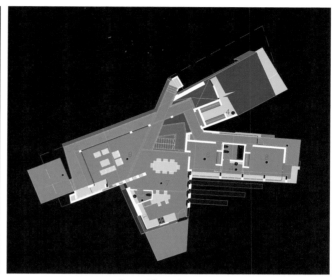

Second floor plan

汉普登路住宅 Hampden Lane House

Location Maryland, USA
Size 2,200 sq ft
Architecture design Robert
M. Gurney
Project design Brian Tuskey
Photographer Maxwell
MacKenzie
Owner Wittheld

The client for this project was a young, forward thinking entrepreneur with no desire for a nostalgic or revivalist style house. He desired a house that was efficient with a minimal footprint, leaving the majority of the lot unoccupied by building and hardscape. The close proximity to a more urban downtown Bethesda warranted a house designed with closer ties to an urban area than to the rural countryside that once informed the design of houses built in Edgemoore.

After much deliberation, it was decided to remove an existing inefficient structure and replace it with a new one. The new house occupies one third less area than the original structure and is sited to maximize green area on the property. Designed as a cube, the new house is approximately 2,200 square feet with no unused or underutilized spaces. The flat roof provides an additional 1,100 square feet of outdoor living space with views of treetops and the downtown Bethesda skyline. Fenestration in the ground faced block walls, composed of varying sized rectangular and square openings, is arranged to optimize views to the green spaces while minimizing views of adjacent houses in close proximity. A series

of landscape walls orchestrate the relationship between the street, required parking court and house. Interior spaces are open and light filled with crisp detailing. Walnut flooring provides a rich base for white walls and millwork, designed in juxtaposition to the charcoal gray exterior walls.

This house represents a deliberate departure in both the thought process and the realization of current building trends in the neighborhood. Instead of building a large house with pretentious ties to the rural past, this new house is smaller with a stronger relationship to the modern urban area that Bethesda has become. The house is intended to be more site sensitive, environmentally conscious, and to provide comfortable, efficient living spaces.

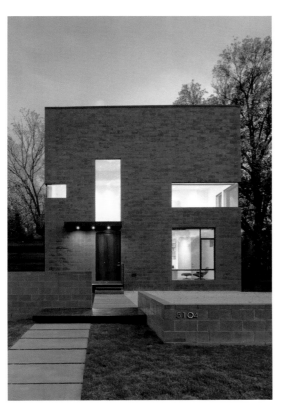

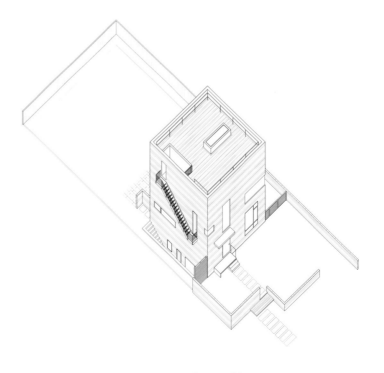

Axonometric

Site plan

First foor plan

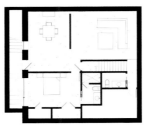

Basement floor plan

Sumaré住宅 **Sumaré House**

Location São Paulo, Brazil
Site area 700㎡
Built area 598.39㎡
Architecture design Isay Weinfeld
Photographer ©2011nk@nelsonkon.com.br
Collaborator Domingos Pascali
Project manager Monica Cappa Santoni
Team Juliana Scalizi, Elisa Canjani, Ilza Fujimura, Marina Cappochi, Juliana Garcia, Leandro Garcia, Gustavo Benthien, Priscila Araújo, Fábio Rudnik
General contractor Anf Engenharia Ltda
Structural engineer Kurkdjian & Fruchtengarten Engenheiros Associados
Mechanical, electrical and plumbing engineer Sphe Engenharia s/c Ltda
Air conditioning engineer Assistec Serviços Técnicos Ltda
Landscape architect André Paoliello

Sumaré house is located in São Paulo, and was designed for a graphic designer. The client wanted a spacious house, where she could work, exercise, entertain friends and, of course, live in. Thus, we had to fit an atelier, a swimming pool and a space for ballet routines into the house, along with ample entertaining areas, two bedrooms and all other rooms suitable to a residence.

The plot is not a small one (700 sq mt), but due to construction (height) restriction laws, the building should not exceed 2 floors, and an underground floor was necessary. There, we placed the caretaker's quarters and the atelier – as both areas open onto small but nice lawns, one does not feel like being underground at all. On the middle floor, a few steps above street level, there are the sitting and dining rooms, the kitchen and a larger lawn. In the living room, we designed a long étagère so the owner could display her collection, ranging from works of art to design and vintage objects. On the upper floor, there are the bedrooms – hers and a guest's – and the "facilities" for physical exercising: ballet and swimming. As this is a space she meant for private use mainly, she liked it when we suggested having it in her bedroom, enclosed only by a screen of pre-cast concrete blocks, through which she would be able to see the city skyline far away. Above all, on the uppermost level, there is a wood-decked outdoor entertaining area, perfect for a get-together, by day or by night.

1. Garage
2. Security Booth
3. Entrance Hall
4. Mechanical Room
5. Laundry Room
6. Caretaker's quarters
7 - Garden
8. Atelier
9. Storage

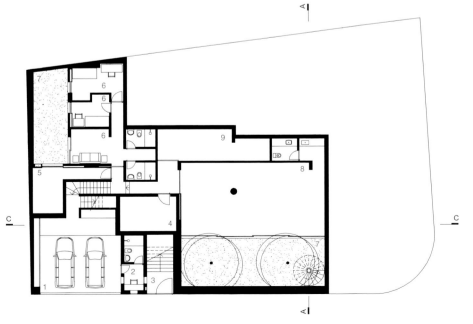

Basement plan

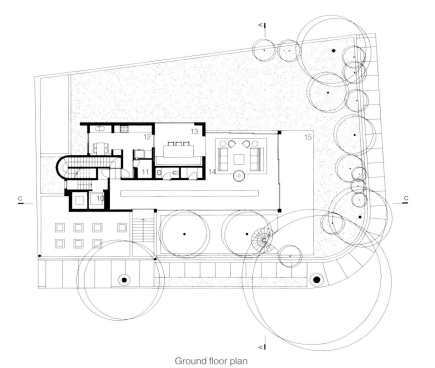

10. Elevator
11. Pantry
12. Kitchen
13. Dining Room
14. Living Room
15. Veranda

Ground floor plan

Hover住宅（系列3） Hover House 3

Location Venice Canals,
LA, USA
Bldg. area 2,500 sf
Architecture design Glen
Irani Architects
Photographer Derek Rath

Situated on the Venice Canals of Los Angeles, California, Hover House 3 represents the third in the architects' Hover House series. This series focuses on maximizing outdoor living on small lots by 'hovering' the building envelope above the grade level in order to create space for outdoor living environments. This series proposes that interior living space be reduced in favor of less resource-intensive outdoor living amenities. As material and labor costs increase in the coming decades, increasing outdoor functionality while decreasing indoor area in temperate climate zones is one solution to the rising cost and over-consumption of building resources.

While this 3-bedroom, 2-office, 2500 SF house already represents a substantial reduction in indoor floor area (about 25% from the norm), the inhabitants of this and two other Hover Houses (including the architect's own house) enable us to study the effectiveness of this model and refine an approach to suit mainstream culture.

Hover House 3 responds to the tight confines of its 32',95' lot on the Venice Canals with little pretense as a simple box elevated over the landscape that is fully programmed to facilitate all the functions of a living room, dining room and kitchen. The interior program for the same functions (which the client unfortunately could not be convinced to

substantially forego) was reduced in floor area by over 50%.
The hope is that the Hover House concept takes root in the community as a practical model for exchanging built volumes with exterior living equivalents. Ultimately, the homeowners will dictate if the Hover House model can actually exclude the interior community areas to some degree, thus saving cost, resources and reducing the carbon footprint.

Hover House 3D Section

1. Library
2. Study
3. Bath
4. Covered Patio
5. Coat Closet
6. Kitchen
7. Stair
8. Living/Dining Area

First floor plan

1. Bath 02
2. Bedroom 03
3. Bedroom 02
4. Laundry
5. Master Closet
6. Master Bedroom 01
7. Master Bath

Second floor plan

条形住宅 Barcode House

Location Washington DC, USA
Site area 87㎡
Addition area 22.3㎡
Architecture design David Jameson Architect
Design David Jameson, FAIA, Alex Stitt, project architect
Contractor The Ley Group
Photographer Paul Warchol Photography

Barcode House explores juxtapositions between the heavy and the light, the old and the new. The work is formed by positioning the project's diverse pressures into a unique situational aesthetic. Brittle masonry walls of the existing Washington, DC row house governed that the addition be engineered as a freestanding structure. Site constraints dictated a vertically oriented spatial solution. The client's desire for transparent living space generated the opportunity to create an integrated solution for lateral force requirements. Structural steel rods within a glass window wall are aligned with datum lines of the neighboring building elevations. A stucco circulation tower anchors the living space to the existing row house.

1. Sitting Room
2. Kitchen
3. Vestibule
4. Stairway

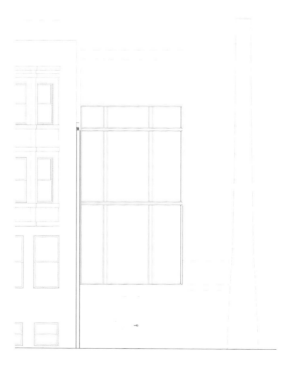

South elevation

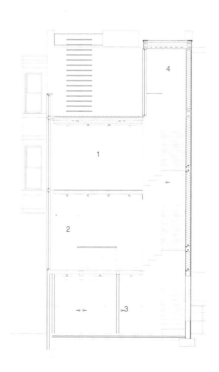

Section

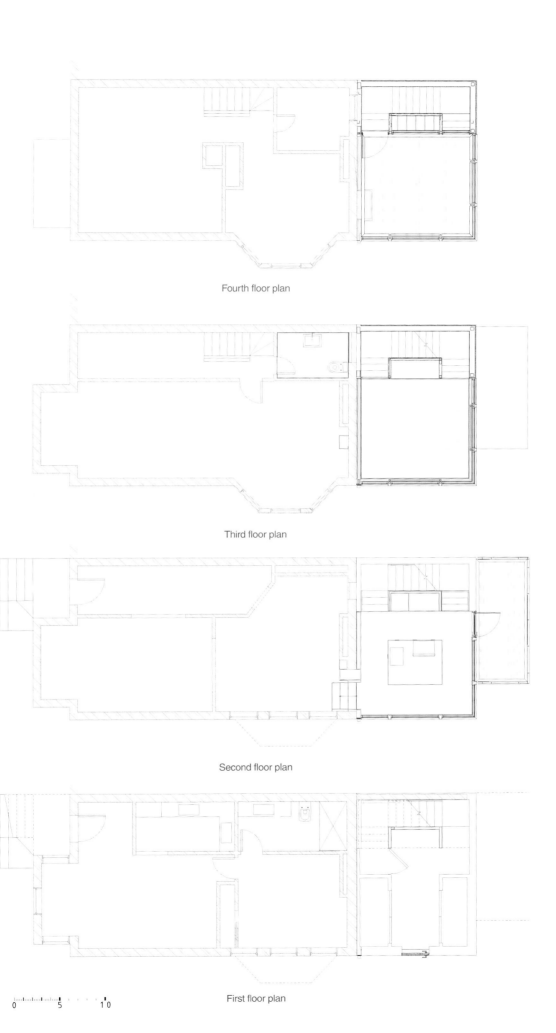

Fourth floor plan

Third floor plan

Second floor plan

First floor plan

0   5   1 0

湖景环保住宅 Mary Lake Residence

Location Ontario, Canada
Size 3,000 sf
Architecture design Altius
Design participation Trevor McIvor,
Tony Round, Logan Amos
Engineering Hamann Engineering,
Toronto
Photographer Jonathan Savoie

The project is comprised of two solid volumes connected by a glazed entry space and indoor/outdoor living room. The roof of the living room is sheltered between the two solids and becomes an additional outdoor space with a view over the lake, a garden and a hot tub. The home is a balance of program spaces, with the public entertaining spaces of the living room, dining room and kitchen and the more private and intimate spaces of the library and den. This year-round residence takes advantage of the entire site set amongst towering white pines.

1. Room
2. Bath
3. Living
4. Living
5. Dining
6. Kitchen
7. Living

Floor plan

ARC住宅 ARC House

Location Hampton, NY, USA
Architecture design Maziar
Behrooz Architecture
www.mbarchitecture.com
Photographer Matthew Carbone

The Arc House is a private residence designed for a couple and their two big dogs. The arc contains the living, dining and kitchen areas in an open plan. The entry canopy to the arch and a few other components within it are at a height of 7', creating a touchable lower plane within the larger structure. To allow the client flexibility to add future bedrooms, we decided to house them in a flat- roofed section, in the rear, that is made of structural insulated panels (SIPS). The lower level houses an office, a sitting room, a garage and work-out areas; part of the lower level opens to a courtyard that allows cool air circulation and a distinctly private outdoor space. By splicing the house into the landscape, we were able to transform the flat site into one with multiple horizons and take advantage of natural cooling and passive design. The energy usage of the house is 1/5 of conventional houses, due to both the arc shape and materials used. Conventional lumber is only used in interior partition wall studs and the exterior cedar siding around the bedroom section.

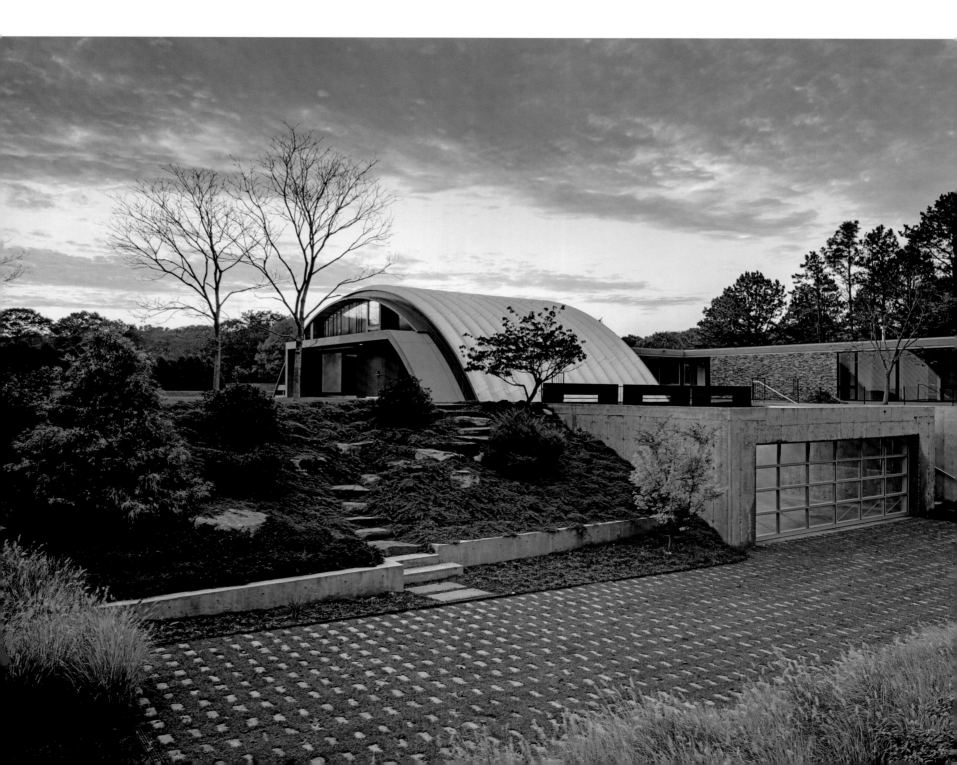

23.2住宅 23.2 House

Location Surrey, British Columbia, Canada
Area 500㎡
Architecture design Omer Arbel Office Inc.
Design participation Omer Arbel, Mark Dennis
Photographer Nic Lehoux

23.2 is a house for a family built on a large rural acreage. There is a gentle slope from east to west and two masses of old growth forest defining two outdoor rooms each with its own distinct ecology and conditions of light; the house is situated at the point of maximum tension in between these two environments, and as such acts at once to define the two as distinct, and also to offer a focused transition between them. The design of the house itself began, as a point of departure, with a depository of one hundred year old Douglas Fir beams reclaimed from a series of burned down warehouses. The beams were of different lengths and cross sectional dimensions, and had astonishing proportions - some as long as 20 meters, some as deep as 90 cm. It was agreed that the

beams were sacred artefacts in their current state and that we would not manipulate them or finish them anyway. Because the beams were of different lengths and sizes, we needed to commit to a geometry that would be able to accommodate the tremendous variety in dimension, while still allowing the possibility of narrating legible spaces. We settled on a triangular geometry.

We folded wood triangular frames made of the reclaimed beams to create roof which would act as a secondary (and habitable) landscape. We draped this artificial landscape over the gentle slope of the site. We manipulated the folds to create implicit and explicit relationships between indoor and outdoor space, such that every interior room had a corresponding exterior room.

We wanted to maximize ambiguity between interior and exterior space. We removed definition of one significant corner of each room by pulling the structure back from the corner itself, using bent steel columns. We introduced large accordion door systems in these open corners so that the entire façade on both sides of each significant corner could retract and completely disappear.

We developed a detail that would allow the beams to define not only the ceilingscape of each interior room, but also to read strongly as elements of the building façade.

木兰公寓 Magnolia Residence

Location Seattle, USA
Bldg. area 2,600 sf
Architecture design Heliotrope
Architects
Photographer Mark Woods
Contractor Dovetail

A new residence located in Seattle's Magnolia neighborhood designed as a simple three-story form wrapped in a cedar rainscreen and stretched across the width of the property to maximize views of west toward Puget Sound and the Olympic Mountains. Spaces are arranged according to the importance of prospect and refuge. Primary social spaces are located on the top floor where the view is the best and where kitchen, dining and living flow seamlessly together in one loft-like space. Private functions – sleeping and bathing – occur on the second level screened from the street. The ground floor contains entry, garage and an opening through the house, front to back – a covered exterior gathering space that frames the view of water and mountain for the passerby at the street.

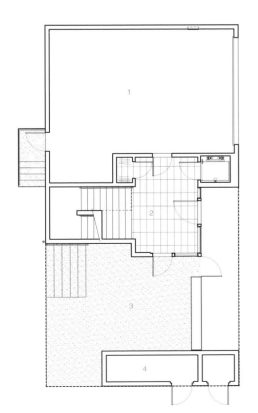
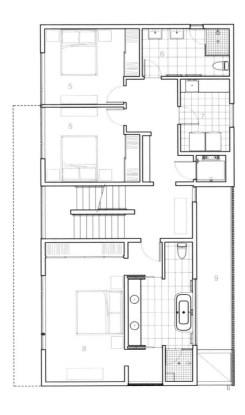
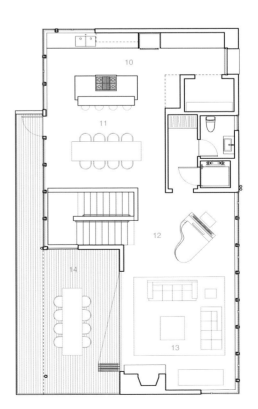

1. Garage
2. Entry
3. Covered Patio
4. Mech
5. Bedroom
6. Bathroom
7. Laundry
8. Master Bedroom
9. Open
10. Kitchen
11. Dining
12. Music
13. Living
14. Deck

First floor plan

Second floor plan

Third floor plan

大洋洲OCEANIA

亨利大街的Barwon Heads住宅
Henry Street, Barwon Heads

Location Victoria, Australia
Size 370m²
Architecture design Jackson
Clements Burrows Pty Ltd
Architects
Design participation Jon Clements,
Graham Burrows, Tim Jackson,
Chris Botterill, Nick James
Contact Person Jon Clements
Landscape design Tim Nicholas
Builder Irene Morgan
Structural Lambert and Rehbein
Phtographer John Gollings

Barwon Heads is in a period of significant change. Heritage overlays currently protect older fishing shacks whilst the less significant built fabric remaining in the seaside town is progressively being redeveloped, and architecture is now significantly contributing to the evolution of this small coastal township. In this case, a young family engaged JCB to design their new permanent residence which would replace a dilapidated 1950's two-storey house that was beyond repair. A sculptural building form emerged from the client's brief which jokingly requested a planetarium as an inclusion. This led to the exploration of circular forms and resulted in a circular skylight in the first floor living areas as a direct reference, however the house was primarily conceived to immerse itself over time as a natural extension of the Ti-tree dominated landscape. The house is wrapped in as skin of vertical cedar battens (providing privacy and solar protection) which appear to emerge from the front fence intentionally confusing the relationship between built form and landscape.

East elevation

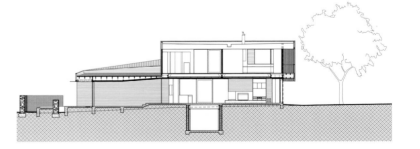

Section

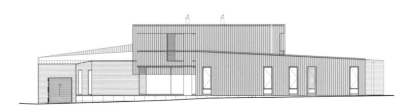

South elevation

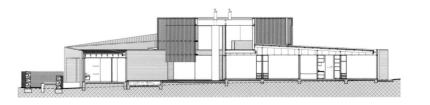

Section

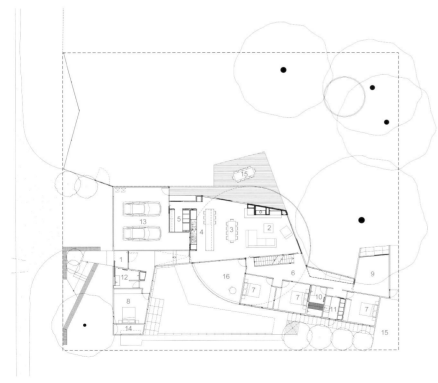

1. Entry
2. Living
3. Dining
4. Kitchen
5. Scullery
6. Hall
7. Bedroom
8. Guest Bedroom
9. Rumpus
10. Bathroom
11. Powder room
12. Laundry
13. Garage
14. PoolStore
15. Deck
16. Two Storey Void

1. Master Bedroom
2. Walk-in Robe
3. Study
4. Retreat
5. Deck
6. Roof Over

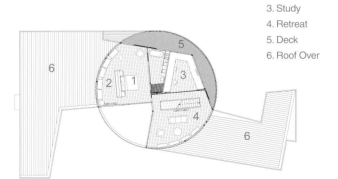

Ground floor plan

First floor plan

Carey住宅 The Carey House

Location Australia
Total planned area 542㎡
Design architect Tony Trobe,
TT Architecture
Project architect/Draftsman
Clinton Atkins
Interior design KMD Design
Structural engineer ADM
Photographer studio2point8
Builder Solve Projects

The Carey house client's brief reflects a 21st century reference to the confidence and optimism that lies at the heart of modernism stating that:"The house is to be a modern, contemporary residence of premium quality using materials not normally used in residential construction – steel, glass, metal cladding etc." In the client's own words "Understated but brave, edgy without being pretentious".

Responding to the brief involved exploring a core idea of one of the modern movement's most enduring legacies; the dissolution of the traditional boundaries between interior and exterior space. The design of the house explores the notion of a primary space being "one room thick". The voluminous central space promotes good cross ventilation and a high degree of hot air "purging" in summer. The design responds to give regionally and seasonally appropriate spaces. In physical form the referencing of Neutra is contained in the notion that the viewer's eye is drawn out by the long horizontal planes of the roof.

The design is not content with just being an exercise in planes and modernism but responds to the contemporary concern with resource sustainability. It makes strong gestures as a standalone, autonomous entity. The following initiatives are taken from concept to completion:
• "Solar slot" orientation
• high thermal mass

- The inclusion of thermal chimney technologies
- The use of a hybrid semi-commercial "whole of building" heating solution including the domestic hot water. The system consists of a solar evacuated tube array that heats up a 1200 litre water vessel from which water is pump through a hydronic slab heating system. The top third of the vessel contains a domestic hot water heat exchange unit from which a domestic hot water ring main runs around the building for almost instantaneous hot water up to 20 metres away from the hot water vessel
- 5,000litre capacity water harvesting system for use for the swimming pool, toilets and gardens. This has allowed an average daily use for a family of 4 of only 400 litres

- Provision for a 2000 litre grey water recycling system
- Provision for a 5KW "back to grid" photovoltaic array
- Installation of thermally improved aluminium, argon filled double glazed window units
- The use of High grade rock wool insulation to provide superior levels of acoustic and thermal insulation including the careful infill around windows and doors to ensure the building to be"air-tight".
- No air-conditioning, six star energy rating

In the resolution of the aesthetic and the technical, house stands as a marriage of modern concerns to modernism gestures.

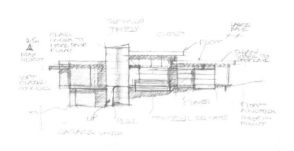

Elevation sketch

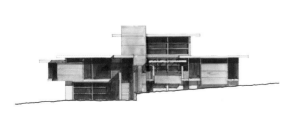

Elevation

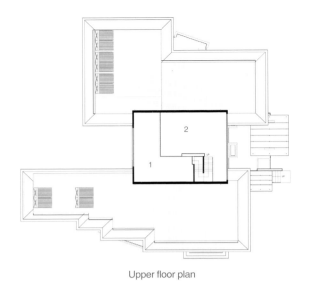

Upper floor plan

1. Loft
2. Family Void
3. Lounge
4. Dining
5. Study
6. Bed 1
7. W.I.R
8. ENS
9. GYM
10. Kitchen
11. Meals
12. Family
13. Rumpus
14. Games
15. Bath
16. Laundry
17. Bed 2
18. Bed 3
19. Bed 4
20. Garage
21. Hobby Room
22. Tanks

Main floor plan

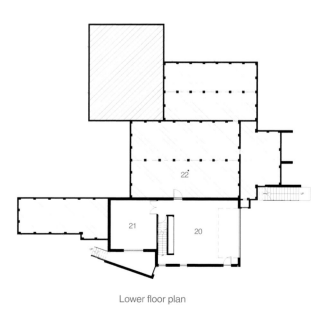

Lower floor plan

Bridgeman Downs住宅 Bridgeman Downs House

Location Bridgeman downs, Australia
Site area 10,000㎡
Bldg. area 1,300㎡
Architecture design Ellivo Architects
Design participation Mason cowle
Interior design Tanya Zealey
Photographer Scott Burrows

This generous family home is arranged in a simple cross form. Each arm of the cross contains distinct functional areas including living, sleeping, entertainment and work zones. The cross form in turn forms four external courtyards also with their specific functions including children's grass play area, pool entertainment zones and tennis court. This simple arrangement of pavilions allows for single room depth spaces that allow good visual and physical connection between the inside and the outside, and promotes cross ventilation. The master bedroom is situated at the upper level giving a sense of separation and privacy from the more public spaces at ground level. The entry into the home from the porte-cochere is via an elongated colonnade flanked by the black lined lap pool.

This colonnade leads to the double height glazed entry that features a stone wall and timber stairs leading to the master bedroom floating above an internal pond. Contained within the entertainment wing is the double height entertainment area which forms a central focus to the house. Eight-metre-wide bi-fold doors create a seamless connection between this space and the double island bench kitchen area.

The client's association with the concrete industry led to the use of monolithic concrete as the primary material. Gloss black off form walls contrast with sand blasted white concrete columns and polished concrete floors, creating a neutral palate of colours and textures against which more sleek materials

are highlighted. Interior finishes have been selected on how well they complement the concrete structure. By contrast, the master bedroom which is dominated by grey and black gives the space a more intimate quality.

1.Balcony

2.Bed room

3.Living room

4.Dining room

5.Kitchen

6.Pond

7.Children's play room

8.Garage

9.Office

10.Outdoor entertaining

11.Indoor entertaining

12.Guest room

First floor plan

Second floor plan

1. Living room
2. Kitchen
3. Bed room
4. Pond
5. Office

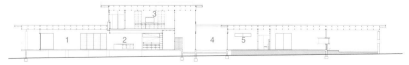

Section

Chelmer住宅 Chelmer House

Location Chelmer, Australia
Site area 1,000㎡
Architecture design Ellivo Architects
Design participation Mason Cowle
Interior design Tanya Zealey
Photographer Scott Burrows

The Chelmer House takes advantage of the dual frontage lot, and was designed as a series of connected spaces and courtyards of varying volumes that run along a central axis. The contrasting high and low volumes are both dramatic and allow natural light, cross ventilation and also define functions of areas. The residence sits within a lush sub-tropical garden, and is a private retreat for the client to entertain or relax. The double height dining space with glass and stainless steel bridge over form the centre-piece of their living areas. Upstairs, across the glass bridge, the master suite is private and overlooks the pool. Internally the surfaces of glass, stone and painted white plasterboard are clean and slick. The timber flooring and soffits add warmth and contrast to the clean palette. Detail in the stair stringers and roughness of the stone feature walls are utilised to provide the textural balance.

1.Garage   6.Dining room
2.Study room   7.Living room
3.Guest room   8.Bed room
4.Media room   9.Master bedroom
5.Kitchen   10.Balcony

Upper floor plan

Lower floor plan

克里夫顿山住宅 Clifton Hill House

Location Melbourne, Australia
Built area 250 m²
Architecture design Sharif Abraham
Architects, www.sharifabraham.com
Photographer Matthew Stanton

The site spans two streets and has two frontages. An original Art Deco house faced the main street. The client's brief was to refurbish this house, providing two new bathrooms and to add an open plan living space to its rear. Our objective was to design an addition that was sympathetic to the original house, whilst advancing the stylistic and spatial qualities of its architectural style.

The addition is a collection of sculptural roof forms oriented to provided outlook and sunlight to the interior as well as responding to the setback regulations of the local building code. Made from concrete with fine stucco finish and aluminium edging, the walls are pierced by large sheets of glass and high windows, establishing transparency and lightness to the form and a sensitive relationship with the future garden.

The interior spaces contrast curved black timber with white walls. Ceilings trace the inside of sculptural roof elements to create double-height spaces and then curve down to create intimate spaces. A linear element replicating the trunk of a tree extends the full width of the room, establishing visual continuity with the kitchen and the garden beyond. The timber (Ebony Macassar) is sourced from the trunk of a single tree. The outside of the trunk, where the grain is younger, is located high in the space and it progressively descends to the joinery and intimate spaces where the core is dense and dark. The consistency of the application of the veneer to the upper portions of the walls is intended to make the living areas feel more intimate whilst maintaining the roof height needed to capture sunlight continuously throughout the day.

Between the addition and the original house, a new courtyard is located to provide light to the inner rooms and living spaces. A discreet staircase at one corner leads to an upper deck located between the roof forms, where panoramic views of treetops over the roof of the original house are achieved.

A corridor with a continuous light beam emanating from its

ceiling, guides the inhabitants away from the addition and into the original house. It leads to the two new bathrooms. The first bathroom is clad in black tiles and presents a dark cavernous experience by surfacing the floor, walls and ceiling with dark textured tiles. The darkness of the interior is intended to focus on the user's (naked) flesh. By contrast, the other bathroom is open to the courtyard and is flooded with natural light. It includes a dressing area with cupboards at one end. Here, continuous bronze tiles unify the space and accentuate the natural light as it moves across their surfaces.

The corridor and the bathrooms are intended as counterpoints to the architecture of the original house, posing questions about the nature of the old and the new and at the same time suggesting possibilities for their reconciliation.

1. Entry
2. Living
3. Kitchen
4. Dining
5. Bedroom
6. Bathroom
7. En-Suite
8. Laundry
9. Hall
10. Deck
11. Cupboards
12. Utility/ Store
13. Void

Ground floor plan

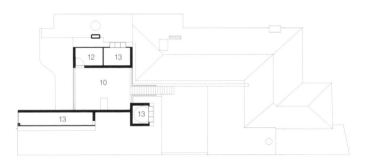

Roof plan

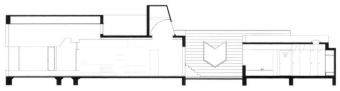

Section C

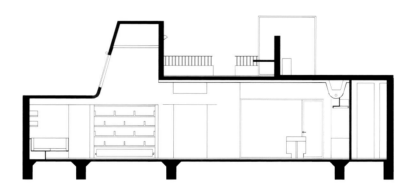

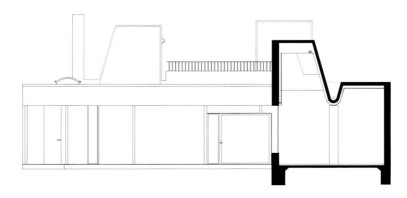

Sections

OCEANIA 大洋洲

Turramurra公园住宅 Turramurra Park House

Location Turramurra, NSW, Australia
Architecture design Liquid Architecture
Photographer Willem Rethmeier

The existing dwelling was a single-storey 1960s brick bungalow elevated above the street, with a cramped plan that under-utilised the opportunities of the sloping site. The clients therefore desired to rework and extend the existing floor plan to create a spacious, open family home with a strong connection to the site and its landscaping. A unique contemporary aesthetic was desired, integrating their own collection of paintings, furniture and objects collected from South-East Asia.

The existing ground floor was simplified to clarify the spatial flow and frame views both internally and to the backyard. The new entry stairs are veiled from the street via a freestanding timber screen panel, creating a transitional indoor/outdoor space folding in towards the main entrance. A second storey was added to maximise the opportunities for living space on the ground floor, with voids connecting the levels visually and spatially, through which skylights channel light into the centre of the home. Bold colour is used to link spaces and enliven the interior, while a new outdoor entertainment area and pool connect the indoor and outdoor spaces. Material is also used to delineate and articulate the composition of volumes in conjunction with the horizontal and vertical elements.

East elevation

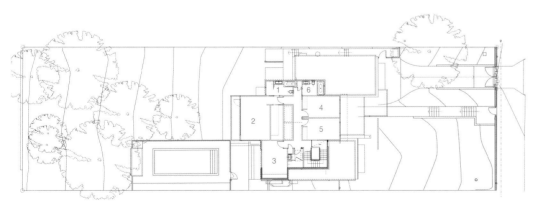

1. Ensuite
2. Bedroom 01
3. Bedroom 02
4. Bedroom 03
5. Bedroom 04
6. Bath

First floor plan

普拉汉的白色住宅 The White House, Prahran

Location Victoria, Australia
Area 362m²
Architecture design Nervegna
Reed Architecture + ph Architects
Design architect Toby Reed
Design participation Toby Reed,
Peter Hogg, Anna Nervegna
Photographer John Gollings

The client, an Art Gallery director, asked for a contemporary home on a narrow inner city allotment. The house was to have two bedrooms plus an extra study that could be used for visiting artists to stay in, and was also to include a private subterranean gallery.

The house extends over 3 levels; the entry is on the middle, ground level, where all the living spaces flow around a courtyard, a sliced circle, a hint of a possible larger courtyard in an expanded field of scattered architectural objects and events. The placement of objects on this level works much like a layout of a pinball machine, with each surface hinting at possible routes that one could take through the house, sometimes encouraging a certain movement, sometimes not. The front

study is for visiting artists and has an adjoining bathroom. Downstairs is a basement gallery, indirectly lit from the north front garden via a concrete light shaft / skylight which also doubles as a seat or sculpture podium.

The house works a bit like a Rorschach test, enabling people to read what they like in it, whether it be virtual images of a "?", or a number "2" in the front facade, or other architectural images throughout the spaces, hidden like vague reflections. The house in some ways became an experiment in treading the fine line between representational imagery and "pure" abstract form. No matter how abstract our forms may be, the free association of the inhabitants will create new, varied and ever changing meanings and images.

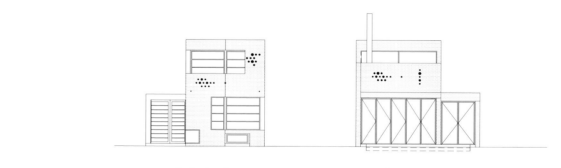

North elevation   South elevation

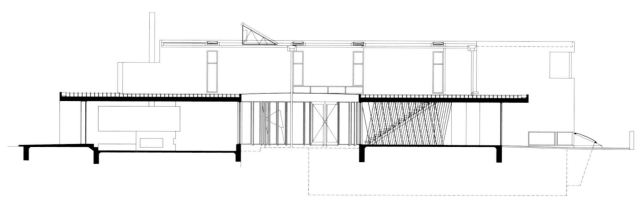

Section

Upper floor plan

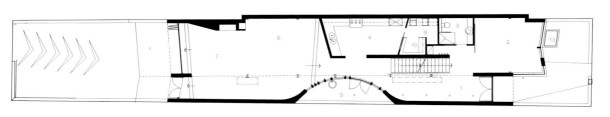

Ground floor plan

1. Entry
2. Guest / Study
3. Bath
4. Kitchen
5. Court
6. Dining
7. Living
8. Terrace
9. Laundry
10. Pantry
11. Robe
12. Ensuite
13. Skylight
14. Master Bedroom
15. Bedroom

欢喜山住宅 Mt. Pleasant Residence

Location Western Australia, Australia
Site area 1,200㎡
Bldg. area 716㎡
Architecture design WrightFeldhusen
Architects
Design participation Tim Wright,
Director
Photographer Robert Frith(Acorn
photo)

The site has spectacular eastern views across the Canning River to the Darling Escarpment. The existing residence was torn down to facilitate a clean site for a family of 5. The clients are enthusiastic entertainers with three young children so the design had to be accommodating for large gatherings and yet remain functional for the family's day-to-day activities.

The site slopes roughly 3m to the river side and therefore facilitated the opportunity for a semi-subterranean entertainment area, while maintaining views to the river. The design intent was to provide transparency of the ground floor living spaces. This means that some areas obtain views by looking through other spaces or courtyards. The more box-like, intimate bedroom areas hover over this glass living zone. The main bedroom projects like a telescope into the view which ensures a dramatic vista, and also provides a shaded eternal entertainment area to the terrace below. All directly exposed windows are protected by operable eternal louver blinds to control solar gain. Passive climate control is the main ESD feature of the house. External walls — especially the west facing bedrooms — are clad in copper which shields the walls in the afternoon and also weather with the classic verdigris. This adds to the natural pallet of materials and compliments the jarrah glue laminated structural columns in the entry and the main stair gallery.

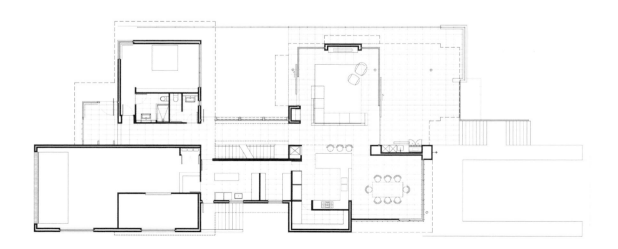

First floor plan

Second floor plan

9 Elmstone住宅 **9 Elmstone**

Location Auckland, New Zealand
Floor area 400㎡
Architecture design Daniel Marshall
Architects, marshall-architect.co.nz
Designers Daniel Marshall, Nick Veint,
Karamia Muller, Mike Hartley, Nick
Sayes
Photographers Emily Andrews,
Ernie Shackles

A suburban context – the site splays out from the end of a quiet Auckland cul-de-sac and slopes steeply away towards the north.

The brief was to provide a family home for a young family. It was important to maintain open ground for the boys to play on and to allow for a vegetable garden. The house was to be as economic as possible considering it's 400m² floor area.

The architectural strategy that developed was to arrange house volumes as densely as possible in such a configuration, which dealt with the steep site, and to ensure that ample ground was left in play for the boys. Spaces are arranged vertically over three levels, each with its own distinct outlook toward the garden and pool. The primary form is enclosed beneath a single sloped roof, which allows variation of floor and ceiling levels to individualise the spatial experience throughout the home.

Industrial materiality, precast concrete walls and simple concrete floors, play against the sophisticated yet subtle detailing of more traditional residential finishes. The placement of concrete elements within the home contributes to a thermal mass to absorb the heat from plentiful northern and eastern sun to regulate internal temperature of the house.

@ Ernie Shackles

@ Emily Andrews

@ Emily Andrews

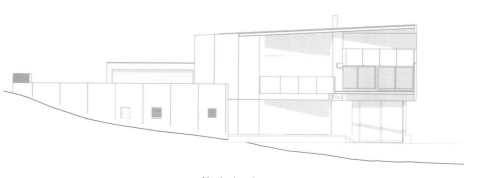

North elevation

West elevation

@ Emily Andrews

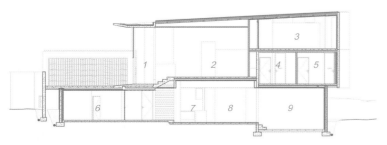

1. Entry Foyer / Gallery
2. Lounge
3. Master Bedroom
4. Kids Bedroom 1
5. Kids Bedroom 2
6. Den
7. Kitchen
8. Dining
9. Living Room

1. Balcony
2. Master Bedroom
3. Ensuite
4. Kids Bedroom 1
5. Kids Bathroom 2
6. Covered Patio
7. Living Room

Longitudinal section

Cross section

1. Pool
2. Lawn
3. Covered Patio
4. Lounge
5. Patio
6. Dining
7. Kitchen
8. Scullery
9. Powder Room
10. Pool Room
11. Den
12. Storage

1. Kids Bedroom 1
2. Kids Bedroom 2
3. Balcony
4. Living
5. Entry Foyer / Gallery
6. Vehicle Court
7. Ensuite
8. Storage 1
9. Storage 2
10. Kids Bathroom
11. Kids Study / Play
12. Storage 3
13. Laundry
14. Garage

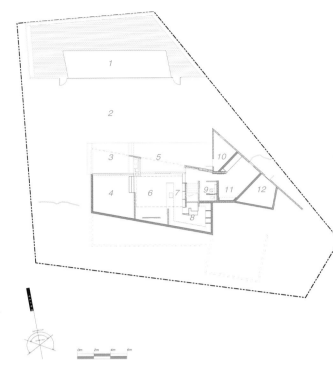

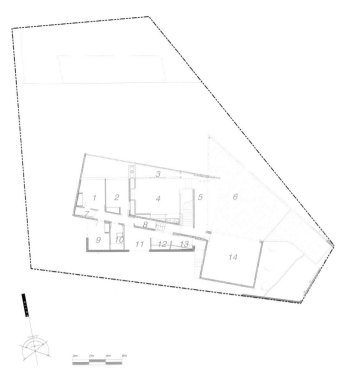

First floor plan

Second floor plan

@ Emily Andrews

@ Emily Andrews

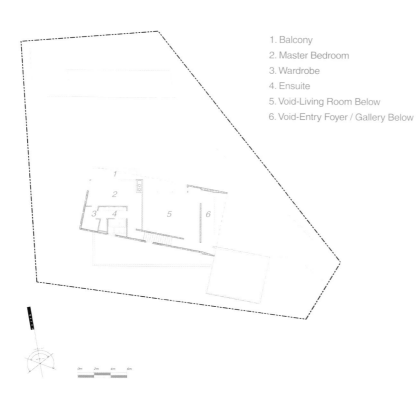

1. Balcony
2. Master Bedroom
3. Wardrobe
4. Ensuite
5. Void-Living Room Below
6. Void-Entry Foyer / Gallery Below

Third floor plan

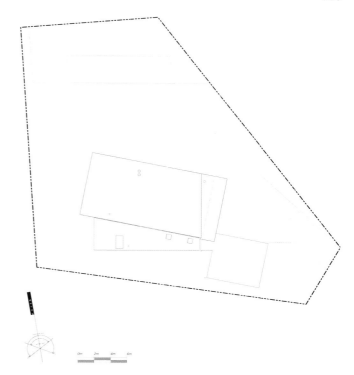

Roof plan

Mash住宅 Mash House

Location Melbourne, Australia
Total floor area 85㎡
Architecture design Andrew Maynard
Architects / Andrew Maynard
Design participation Mark Austin,
Matthew McClurg
Photographer Kevin Hui

The approach taken for the "Mash House" is one which, despite first impressions, celebrates the backyard. Or perhaps less so the traditional notion of the backyard, and more so just plain, outdoor space. The original deep and dark, double-fronted Victorian house offered a plethora of challenges; not least of all, it's lack of solar access. In predictable fashion, services had been attached to the rear of the dwelling over time, effectively dislocating the living areas from the backyard. An old shed, stretching the width of site, sat idly to the rear. These elements combined, meant the overriding feel of the house was one of disconnection.

The young family yearned, above all, a functional living and kitchen space. Squeezing in an ensuite and walk-in wardrobe would be a bonus. Instead of jamming further additions to the rear, a glass walkway pulled from the existing dwelling allows a distinct, spatial break between old and new. The residual space is framed as a courtyard, meaning the new living area has direct access to northern light and associated passive solar gain. In place of the old binary layout of external vs internal, the house is now articulated as three masses — the original dwelling, addition and garage, each punctuated by outside space. A blurred line exists not between old and new, but inside and out. The result is a collection of connected spaces, spoilt for light and air.

Beyond siting measures, every effort has been made to minimise the ecological impact of the extension. Where plausible, new materials have been shunned in favour of their reclaimed counterparts. The entire original dwelling has been re-floored in recycled spotted gum. Where the house sees greater light exposure to the rear, a concrete slab works as an oversized thermal sink. Heat absorbed during the day is radiated throughout the night, helping maintain a constant and pleasant temperature. Ample double glazing and quality insulation throughout furthers this effect. Another significant consideration was that of the carbon footprint; demolition was proposed only where necessary and specifically to break open the stifling internalisation present to the rear of the dwelling.

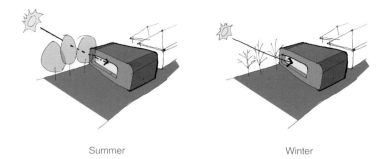

Summer   Winter

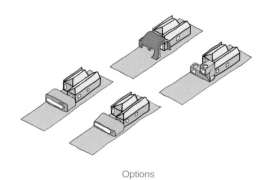

Options

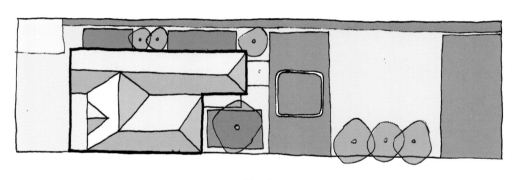

Site plan

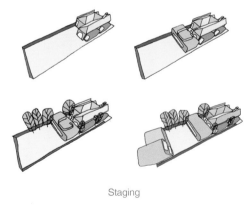

Staging

Bulkhead

1. Kids bedroom
2. Hallway
3. Master bedroom
4. Guest bedroom
5. Ensuite
6. Link
7. Laundry
8. Study
9. Courtyard
10. Pantry
11. Kitchen
12. Living
13. Garden room
14. Garage
15. Bathroom

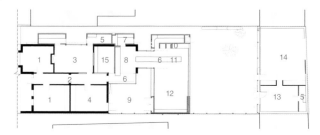

First floor plan

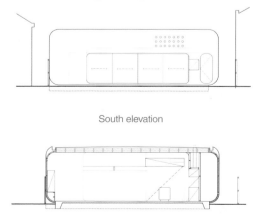

South elevation

Cross section

Oxlade住宅 **Oxlade House**

Location Queensland, Australia
Site area 1,181㎡
Bldg. area 869㎡
Total floor area 940㎡
Architecture design Arkhefield /
Andrew Gutterige
Interior design Arkhefield
Photographer Scott Burrows
(Aperture Photography)

New Farm in Brisbane's inner city has undergone a period of urban renewal, adding a contemporary layer to a suburb noted for its traditional character housing. The Oxlade House is sited within a transitional zone between traditional Queenslanders and a mix of post-1960s housing stock. An eclectic streetscape, town planning codes dictated that the house must adopt a Character Housing typology. This planning constraint challenged and grounded the client's brief for a bold contemporary architecture. Whilst the house has a non-traditional expression, it interprets and abstracts traditional typologies of a solid "core" with integrated verandas, legible roof, and layered screening.

The driving concepts behind the house are "transition" and "orientation" reinforced by the roof as generator of form and space. Abrupt transition creates difference, whereas moments of seamless transition create sameness or openness. The house explores a variety of spatial and experiential transitions from public to private, hard to soft, solid to filter. The landscape and outdoor spaces are interwoven within the navigation and experience of the house. The filtered verandah becomes a buffer to the elevated living room as the first of a linear progression of spaces, from private contained rooms to open, engaged family spaces. The adjacency of the outdoor spaces seamlessly transitions with and expands the family realm. The corners of each room are peeled back to re-orientate attention to the pool, deck and garden.

The roof defines the spatial experience of the house, orienting all spaces to the north east outdoor areas, and capturing service areas along the side boundary. The boldly sloping wall and roof, form an armature to the south responding directly

to overlooking issues, setback requirements and northern orientation. The directional extrusion of the roof adapts to allow functional and contextual contingencies, forcing shifts in proportion, and in turn, exaggerating the roof form.

The dynamism of the roof is an emerging theme in a lineage of Arkhefield's work. Oxlade House continues an exploration into the roof as a subtropical protective device, capturing and defining space. Climatic responses combine with programme and place, informing orientation and transition. These guide a directional language such as extruded eaves, folding, wrapping or capturing of rooms – there is little distinction on "wall". Walls are permeable, invisible or moveable. By drawing from these contextual conditions, a journey ensues in abstraction and invention, often offering unexpected results.

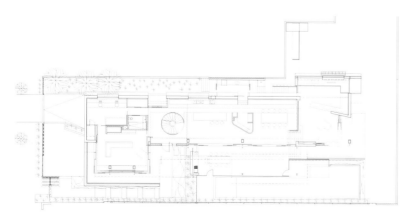

First floor plan

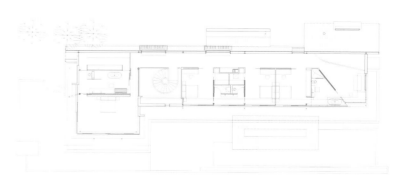

Scond floor plan

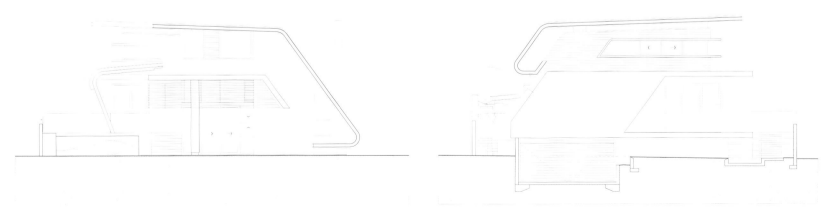

Elevation I   Elevation II

Elevation III

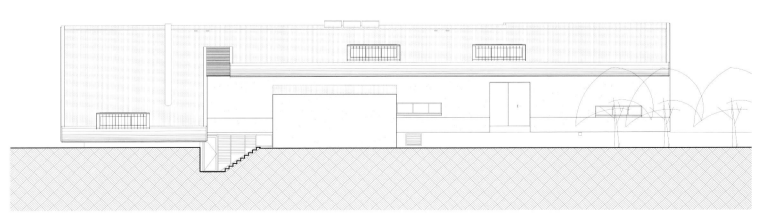

Elevation IV

Split住宅 **Split House**

Location Paddington, NSW, Australia
Site area 197㎡
Floor area 257㎡
Architecture design MCK Architecture
& Interiors (marsh cashman koolloos
architects)
Photographer Willem Rethmeier
Builder Aligra Pty Ltd.
Structural engineer Simpson Design
Associates
Hydraulic engineer Brett Lipscombe +
Associates
Landscape architect Secret Gardens
of Sydney
Surveyor Eric Scerri + Associates
Planning consultant Mersonn
Heritage consultant City Plan Heritage

A relatively modern split level terrace house was refurbished
and enhanced by taking advantage of some site specific
opportunities.
A series of basement rooms were adapted and made functional
which then allowed the internal spaces to connect to the
natural light and ventilation to the south through the creation of
a voided court below street level.

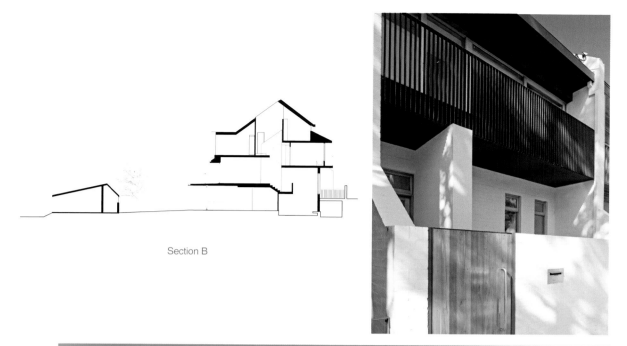

Section B

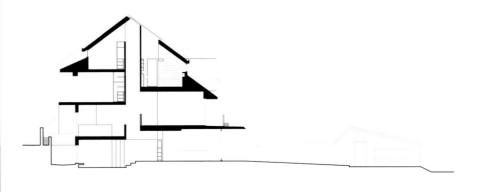

Section A

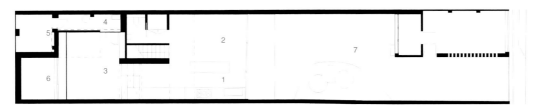

Lower ground floor plan

1. Kitchen
2. Informal Dining
3. Formal Dining
4. Plant Room
5. Basement
6. Courtyard
7. Rear Yard
8. Laundry
9. Garage
10. Bedroom 1
11. Bedroom 2
12. Garden Below
13. Garage + Laundry Below
14. Entry Courtyard Below
15. Courtyard Below
16. Void
17. Pebble Garden Below

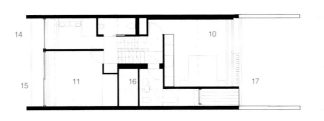

First floor plan

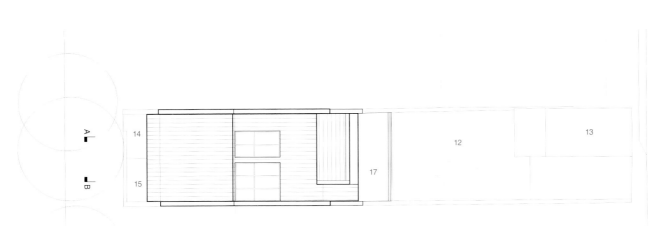

Site plan

山端生态住宅 Hill End Ecohouse

Location Queensland, Australia
Site area 638㎡
Total floor area 261㎡(internal spaces),
52㎡(covered outdoor living), 73㎡
(plant, storage and car & bike & kayak
garage)
Architecture design Riddel Architecture
/ Emma Scragg, Davide Gole
Interior design Riddel Architecture
Photographer Christopher Frederick
Jones

The Hill End Ecohouse is a new house with sustainability at the
core of the design brief. Energy generation and conservation,
water collection, reticulation and recycling, recycled content
in construction materials; low toxicity, durability and low
maintenance are all strong features of the project. Design
element, materials and products have undergone rigorous
assessment of their environmental, social and economic
sustainability credentials. The most unique feature of this house
is its high recycled content — estimated at around 80%.
Design doesn't stop at the outside walls. The surrounding
landscape is designed to provide food for residents and wildlife
and control climatic conditions(sun, summer breezes and winter
winds) to enhance the passive heating and cooling of the house.

East elevation

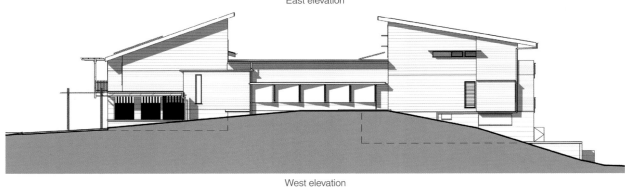

West elevation

1. Bedroom
2. Entrance
3. Gallery
4. Bathroom
5. Gymnasium
6. Terrace
7. Living room

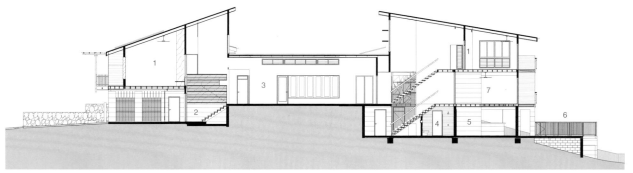

Longitudinal section

1. Garage
2. Entrance
3. Plant room
4. Cellar
5. Media room
6. Hall
7. Bedroom
8. Pool terrace
9. Gymnasium
10. Pool
11. Garden

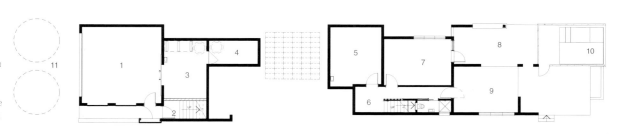

First floor plan

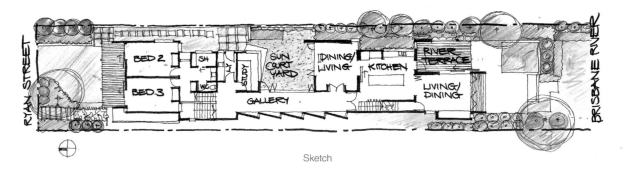

Sketch

北邦迪某住宅 North Bondi House

Location North Bondi, NSW,
Australia
Site area 237㎡
Floor area 180㎡
Architecture design MCK
Architecture & Interiors (marsh
cashman koolloos architects)
Photographer Douglas Frost
Builder Artechne
Structural engineer Simpson Design
Associates
Hydraulic engineer Steve Paul +
PartnersQuantity
Surveyor QS Plus, Eric Scerri

This project comprises a first floor addition and a ground floor reinvention to a space + light house.

Floorspace has been sacrificed on the first floor to create a double height open core to the centre of the house, filtering natural light into the depths of the interior.

Dead space has been otherwise eliminated where possible through open planning and recycling circulation functionality.

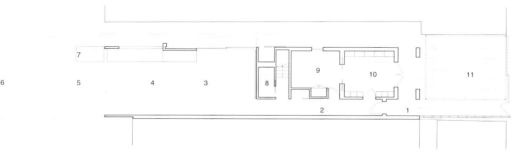

Ground plan

First floor plan

1. Entry
2. Hall
3. Living
4. Kitchen/Dining
5. Terrace
6. Garden
7. BBQ
8. WC
9. Bedroom
10. Study
11. Carport
12. Hall
13. Wardrobe
14. Master Bedroom
15. Deck
16. Ensuite
17. Bedroom
18. Void
19. Bathroom
20. Bedroom

Ilma Grove住宅 Ilma Grove

Location Northcote, VIC, Australia
Architecture design Andrew Maynard
Architects
Design participation Andrew Maynard,
Mark Austin, Matthew McClurg, Cara
Wiseman
Photographer **Kevin Hui**

The Ilma Grove house is Andrew Maynard Architect's greenest house (so far). Its planning and orientation is based solidly around passive solar efficiency. All roof captured rain water is harvested. It has solar panels, high performance insulation, recycled materials, Low-e coated double glazing, low organic VOC materials and, most importantly, it is small.

Ilma Grove house is an extension to a heritage home in Northcote, Victoria. The extension provides a lot more than just additional space to the current home for 3, it provides the environment for a sustainable lifestyle. From the beginning of the design process, the idea of a sustainable home was part of the brief. The client wanted a place which could be practical and could help their family reduce their ecological footprint.

Brick lean-tos at the rear of the original house were demolished

limiting the adaptation of old part of the house & containing it within the hipped roof. This allowed the extension to sit next to the original structure rather than invading the rear of the existing building, avoiding extra costs in demolition and reducing construction waste. Adhering to the principal of "small is green" (less waste, less electricity consumption, less materials, less cost) the result is a functional open plan, where maximizing passive solar gain becomes indispensable. A master bedroom has been included on the first floor with a roof terrace above that overlooks the city to the south and the Dandenong ranges to the east.

Andrew Maynard Architects took advantage of the north facing backyard, and developed an exploration of mass where segments were carved out in order to maximize sun penetration. This generated a geometrical structure where the

internal flesh of the box is revealed with rich timber surfaces, contrasting the raw recycled brickwork.

The interaction between levels and the idea of blurring lines between new and old, inside and out, introduced the idea of integrating the backyard into the interior of the house, carving the garden inside, which currently is being used to grow tomatoes.

The internal flooring is locally sourced bluestone with the intention that using locally produced materials shortens transport distances, thus reducing CO_2. The dark, dense nature of bluestone acts as a thermal mass soaking up the low winter sun & passively heating the house. In summer the high sun does not come into contact with the bluestone allowing the floor to act as a cooling mass in the hotter months. Low-e coated double glazed throughout the house ensures that heat is retained in winter and reduces the penetration of heat in summer. These design features completely remove the need for air-conditioning and drastically reduce the necessity of heating throughout the colder months.

The choice of materials was a vital step in order to create a sustainable structure. It was decided to re-use/re-assemble the existing bricks from the demolished areas of the old part of the house to form the new addition, blurring the line between what is new and what is old. Using recycled materials is a sustainable choice, however there is still a carbon debt accrued by the transport and reworking of materials. The Ilma Grove house avoids this by reusing the bricks of the demolished lean-to on site; what was demolished has been rebuilt in a new configuration. This not only avoids waste, landfill and transportation of materials, furthermore it ties the material language of the new structure back into the original house. Face brick masonry is also a durable and a low maintenance material, which can potentially be recycled again. Reinforcing the thermal performance of the recycled brick is high performance insulation that has been installed throughout the home.

After lengthy discussions about the brief the client made a conscious choice not to have an ensuite upstairs. This helped to reduce the size of the addition while also reducing the embodied energy that comes with doubling up on functions and equipment within the home.

Solar panels have been added to make the house coal independent. Ample solar energy is harvested in winter, while a surplus of energy is fed back into the power grid in summer.

The temptation on a large block is to make a large home. This has been resisted. Maximizing the outdoor space and connecting with it so that it has become a natural extension of the living space was the key. A small house is a sustainable house.

And like our Tattoo House we decided to graffiti stencil our new creation before anyone else got a chance.

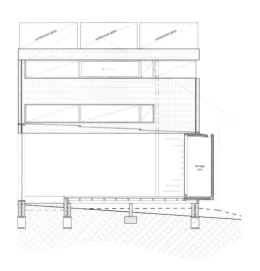

Section I

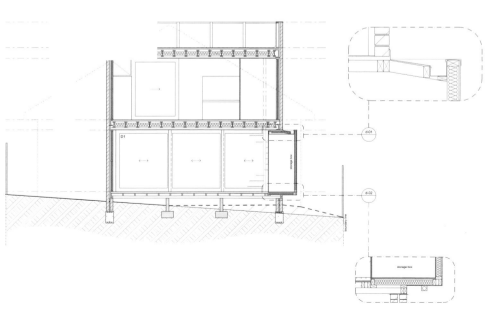

Section II

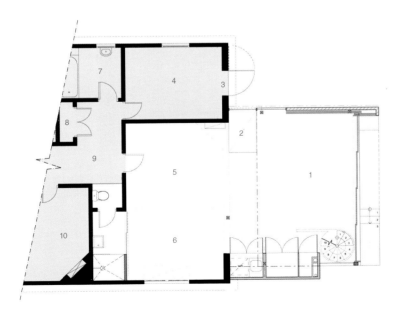

1. Lounge
2. Internal Garden
3. New Entered Opening
4. Bedroom 2
5. Dining
6. Kitchen
7. Bathroom
8. Store
9. Unchanged
10. Living unchanged

Ground floor plan

Doonella Noosa住宅 **Doonella Noosa|House**

Location Queensland, Australia
Site area 647㎡
Total floor area 336㎡
Architecture design Ellivo Architects /
Scott Whiteoak
Design participation Daniel Volpato
Interior design Tanya Zealey
Photographer Scott Burrows

The 647㎡ allotment, on the corner of Jacksonia Place and Morinda Circuit, was chosen because of its north orientation and close proximity to the parkland reserve and Lake Doonella. The house has been designed to maximize the north aspect whilst addressing the two street frontages of the corner allotment. The use of skillion roofs allow for raking ceilings to living areas and the master bedroom. A clearstory running the length of the house invites natural light to enter the centre of the dwelling whilst allowing hot rising air to escape. Attention has been paid to the rear deck area in order to maximize the potential views to the parkland. This also creates privacy to the deck and master bedroom from Morinda Circuit.

The home incorporates design principles that consider the needs of people of varying age and ability. The downstairs bedroom designs are suitable for persons of limited mobility. Push button switches have also been placed at accessible heights to facilitate ease of use.

This house has been designed to meet an exclusive market position in one of the last available land parcels in Noosa. The sensitivity of the environment is recognised and reflected in the specific site water sensitive urban design, civil works, building design and materials chosen. The features are all commercially available and provide viable alternatives to current models of practise within Southeast Queensland.

North elevation

South elevation

West elevation

East elevation

1. Garage
2. Entry
3. Study & Guest room
4. Media room
5. Dining room
6. Deck
7. Living room
8. Bedroom
9. Clearstory

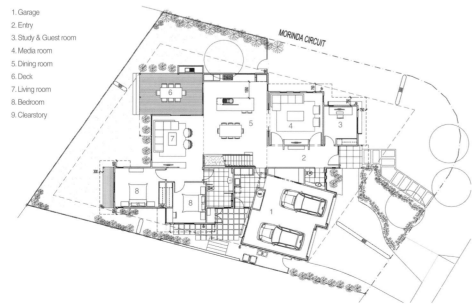

First floor plan

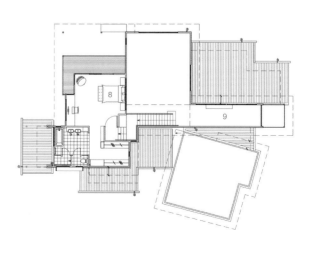

Second floor plan

穿孔住宅 Perforated House

Location Victoria, Australia
Architecture design KAVELLARIS
URBAN DESIGN (KUD)
Client KUD
Photographer Peter Bennetts

The built form is essentially an urban infill within a very small 5.5x14.4m envelope. The perforated house is our proposal to establish an alternative language of our cultural attitudes towards critical questions of identity and heritage. We were interested in retaining the "idea" and the "symbolism" of the terrace but elevating the gesture to an ironic, even satirical, level to incite a public debate. The irony being that only in the absence of matter, through perforation, rather than through a physical reproduction of a terrace house, is the symbol of a terrace house apparent.

We wanted the house to be more than just a facade, more than just a message or a graphic stuck to a building. Our building is not an urban canvas paying tribute to Venturi's "decorated shed". Instead, the external facade can also be experienced internally, as it is a multi-functional device that constantly transforms the building from solid to void, from private to public and from opaque to translucent. By day, the building is heavy and reflective but by night inverts to a soft, translucent, permeable light box. The operable wall, or the absence of the facade, enabled us to critique the idea that houses are static.

The use of operable walls, doors, curtains and glass walls enables the occupants to control the experience and environment. This architectural manipulation of space allows a blurring of the boundaries between inside

and outside, public and private realms. The manipulated spaces overlap and borrow the amenity and context of the surrounding environment.

The plan inverts the traditional terrace program with the active living zones on the first floor opening onto a north facing terrace, thereby generating a primarily northern orientation to a south facing block. The perforated house incorporates passive sustainable interventions such as northern oriented glass bi-fold doors and louvers for cross ventilation as the primary means of cooling. In addition, solar hot water and 5-star rated sanitary fixtures are incorporated.

Reinforced by the childlike mural reminiscent of a past era, the north facing terrace redefines the "Aussie" backyard, and forms a commentary on the changing demography of the family unit and, ultimately, the typology of the inner city house.

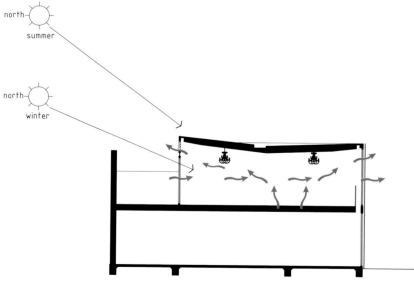

Section

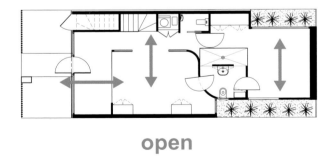

open

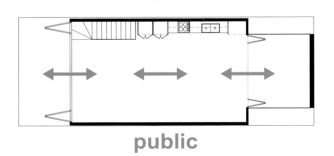

public

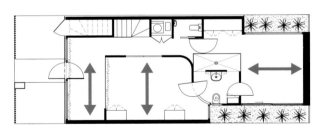

closed

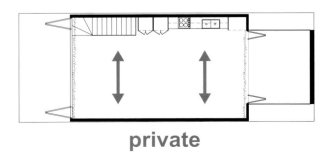

private

1. Entry Foyer
2. Study
3. Bedroom
4. Ensuite
5. Powder Room
6. Courtyard
7. Store

10. Living
11. Kitchen
12. Dining
13. Backyard

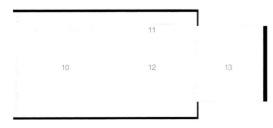
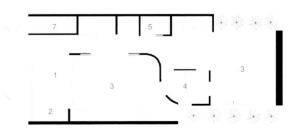

Floor plan

N

Rischbieth住宅 **Rischbieth House**

Location Australia
Design Tony Trobe
Design participation Karin McNamara
Architecture design TT Architecture
Documentation Phil Dale
Interior design Karin McNamara (KMD)
Photography Studio 2point 8 Tony Trobe

The conceptual framework for this dual occupancy in Deakin, directly behind the Prime Minister's residence, flows from the physical considerations of a corner block location and the client's requirement for a very liveable and "retirement friendly" home.

The kernel of the scheme is the idea that the main living space is effectively as "one-room thick". This idea allows for a strong relationship between font and back and from the everyday areas to the sunny northern pocket. This dynamic also permits access to the more private summer courtyard whilst promoting good cross ventilation. This is a consideration often underestimated in the Canberra climate. In order to enable good sun penetration into the family room in colder months and enclose dramatic volumes in the central space the architect has introduced a bold, visually arresting play of curving and floating roof elements.

The demolition of the old poorly oriented original house has enabled a new higher density development to meet the sustainability objectives of the Canberra Territory Plan (i.e. to reduce urban sprawl). The use of a rich and light palette of materials and strong architectural forms is in robust contrast to the conventionality adjacent buildings. The project aspires to catalyse the revivification of one of the more venerable areas of the ACT.

The owners encouraged the architect to "tread boldly" and was pleased with the "on-time, on budget", sustainable outcome. All associated with the project are justifiably proud of the result.

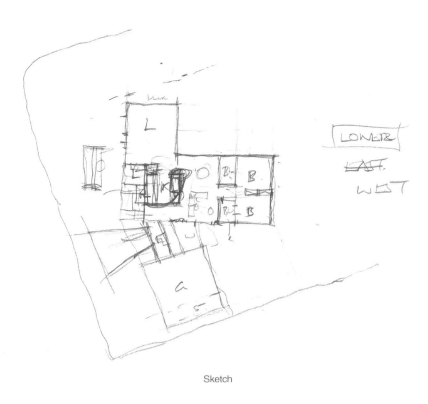

Sketch

Context models

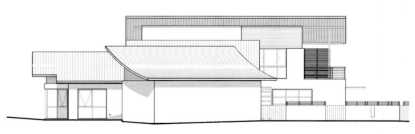

Elevation

1. Lounge/Dining
2. Cellar
3. Powder
4. Pantry
5. Kitchen
6. Meals/Family
7. Bedroom 02
8. Walk-in
9. Bath
10. Laundry
11. Bedroom 03
12. Pond
13. Entry
14. Garage
15. Courtyard

Ground floor plan

Injidup海滨住宅 Injidup Beach Residence

Location Western Australia, Australia
Site area 10,000㎡
Bldg. area 507㎡
Total floor area 602㎡
Architecture design WrightFeldhusen Architects
Design participation Tim Wright, Director
Interior Design WrightFeldhusen Architects
Photographer Patrick Bingham-Hall

The exposed spectacular site has vast panoramic views of Injidup Beach, which is located in the southwest corner of Western Australia. The site is native bushland on an exposed ocean front hill. This natural and visually sensitive site resulted in the local council imposing stringent building bulk and height guidelines; no portion of the building was to be higher then 4m above natural ground level. This therefore resulted in an open, single level residence.

The client required holiday accommodation for their extended family and as such, the sleeping requirements necessitated several bedrooms in separate zones to isolate young children from shared public areas. The site is exposed with the prevailing southwesterly breeze and hot afternoon sun setting into the main view. The house is devised as two intimate sleeping wings that meet at the glazed pavilion that is the main living area. The roof hovers over this area and folds down to the western horizon to provide afternoon sun protection. This roof shelters the main living area and is architecturally articulated to evoke a tent like experience. The main external living areas are focused on the northeast courtyard, with borrowed views of the ocean through the glazed living area. This area is also protected from the breeze. All building materials are included to deal with the harsh marine environment.

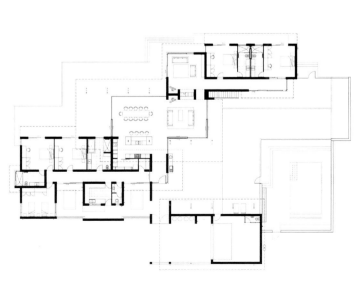

First floor plan

图书在版编目(CIP)数据

155个居住设计：英文/韩国建筑世界出版社编
. —大连：大连理工大学出版社，2012.2
ISBN 978-7-5611-6629-1

Ⅰ.①1… Ⅱ.①韩… Ⅲ.①住宅－建筑设计－世界
－图集 Ⅳ.①TU241-64

中国版本图书馆CIP数据核字（2011）第238681号

出版发行：大连理工大学出版社
　　　　　（地址：大连市软件园路 80 号　　邮编：116023）
印　　刷：精一印刷（深圳）有限公司
幅面尺寸：260mm×300mm
印　　张：58
出版时间：2012 年 2 月第 1 版
印刷时间：2012 年 2 月第 1 次印刷
出 版 人：金英伟
统　　筹：房　磊
责任编辑：张昕焱
封面设计：王志峰
责任校对：王单单

书　　号：ISBN 978-7-5611-6629-1
定　　价：748.00 元（共 2 册）

发　行：0411-84708842
传　真：0411-84701466
E-mail: a_detail@dutp.cn
URL: http://www.dutp.cn